T0292575

Studies in Computational Intelligence

Volume 612

Series editor

Janusz Kacprzyk, Polish Academy of Sciences, Warsaw, Poland
e-mail: kacprzyk@ibspan.waw.pl

About this Series

The series "Studies in Computational Intelligence" (SCI) publishes new developments and advances in the various areas of computational intelligence—quickly and with a high quality. The intent is to cover the theory, applications, and design methods of computational intelligence, as embedded in the fields of engineering, computer science, physics and life sciences, as well as the methodologies behind them. The series contains monographs, lecture notes and edited volumes in computational intelligence spanning the areas of neural networks, connectionist systems, genetic algorithms, evolutionary computation, artificial intelligence, cellular automata, self-organizing systems, soft computing, fuzzy systems, and hybrid intelligent systems. Of particular value to both the contributors and the readership are the short publication timeframe and the worldwide distribution, which enable both wide and rapid dissemination of research output.

More information about this series at http://www.springer.com/series/7092

Roger Lee

Editor

Software Engineering, Artificial Intelligence, Networking and Parallel/Distributed Computing 2015

 Springer

Editor
Roger Lee
Software Engineering and Information
  Technology Institute
Central Michigan University
Mount Pleasant, MI
USA

ISSN 1860-949X   ISSN 1860-9503   (electronic)
Studies in Computational Intelligence
ISBN 978-3-319-23508-0   ISBN 978-3-319-23509-7   (eBook)
DOI 10.1007/978-3-319-23509-7

Library of Congress Control Number: 2015948719

Springer Cham Heidelberg New York Dordrecht London

Printed on acid-free paper

Springer International Publishing AG Switzerland is part of Springer Science+Business Media
(www.springer.com)

Foreword

The purpose of the 16th IEEE/ACIS International Conference on Software Engineering, Artificial Intelligence, Networking and Parallel/Distributed Computing (SNPD 2015) held on June 1–3, 2015 in Takamatsu, Japan, is aimed at bringing together researchers and scientists, businessmen and entrepreneurs, teachers and students to discuss the numerous fields of computer science, and to share ideas and information in a meaningful way. This publication captures 17 of the conference's most promising papers, and we impatiently await the important contributions that we know these authors will bring to the field.

In chapter "On the Accelerated Convergence of Genetic Algorithm Using GPU Parallel Operations", Cheng-Chieh Li, Jung-Chun Liu, Chu-Hsing Lin, and Winston Lo propose to accelerate the evolution speed of the genetic algorithm by parallel computing, and optimize parallel genetic algorithms by methods such as the island model.

In chapter "A GPU-Based Pencil Beam Algorithm for Dose Calculations in Proton Radiation Therapy", Georgios Kalantzis, Theodora Leventouri, Hidenobu Tachibana and Charles Shang conduct studies on Pencil-beam dose calculation algorithms for pro-ton therapy that have been widely utilized in clinical routine for treatment planning purposes in most clinical settings, due to their simplicity of calculation scheme and acceptable accuracy. The studies indicated a maximum speedup factor of ~ 127 in a homogeneous phantom.

In chapter "Incremental Max-Margin Learning for Semi-Supervised Multi-Class Problem", Taocheng Hu and Jinhui Yu propose an incremental max-margin model for semi-supervised multi-classification learning, where efficiency and accuracy need to be considered. Their approach captures essence of the exploration–exploitation tradeoff.

In chapter "Improving Hypervisor Based SSD Caching with Logically Partitioned Blocks and Scanning in Cloud Environment", Hee Jung Park, Kyung Tae Kim, Byungjun Lee, Rhee Man Kil and Hee Yong Youn propose a novel hypervisor-based SSD caching scheme, employing a new metric to accurately determine the demand on SSD cache space of each VM. Computer simulation confirms that it substantially improves the accuracy of cache space allocation compared to the existing schemes. It also allows to display comparable hit ratio as the existing schemes with less amount of SSD cache for the VMs.

In chapter "Emotional Scene Retrieval from Lifelog Videos Using Evolutionary Feature Creation", Hiroki Nomiya and Teruhisa Hochin propose an emotional scene retrieval framework for the purpose of promoting the utilization of a large amount of lifelog videos. The proposed method is evaluated through an emotional scene detection experiment using a lifelog video dataset containing spontaneous facial expressions.

In chapter "On Solving the Container Problem in a Hypercube with Bit Constraint", Antoine Bossard and Keiichi Kaneko propose a routing algorithm selecting in a hypercube internally node-disjoint paths between any two nodes, and such that the selected paths all satisfy a given bit constraint. The correctness of the proposed algorithm is formally established and empirical evaluation is conducted to inspect the algorithm's practical behaviour.

In chapter "Algorithms for Removing Node Overlaps with Some Basis Nodes", Noboru Abe, Hiroaki Oh, and Kouhei Inoue propose three heuristic algorithms to remove node overlaps in graphs with several tens of nodes by refining a previously proposed algorithm, i.e., the force-transfer algorithm.

In chapter "Significant Frequency Range of Brain Wave Signals for Authentication", Preecha Tangkraingkij discusses a new biometric system using brain wave signals (EEG). The purpose of this study is to explore which frequency range of brain wave signals can be utilized for authentication.

In "Simple Models Characterizing the Cell Dwell Time with a Log-Normal Distribution", Naoshi Sakamoto presents two simple models in order to estimate the probabilistic distribution of the cell dwell time. They show that the probabilistic distribution of the cell dwell time of each model is approximated by a log-normal distribution.

In chapter "A Method of Ridge Detection in Triangular Dissections Generated by Homogeneous Rectangular Dissections", Koichi Anada, Taiyou Kikuchi, Shinji Koka, Youzou Miyadera and Takeo Yaku discuss a method for detection of ridges in 3D terrain maps. They introduce the steepest ascent method in triangular dissections generated by homogeneous rectangular dissections.

In chapter "Architecture for Wide Area Appliance Management", Arata Koike, and Ryota Ishibashi studied architecture for Internet-of-Things (IoT) appliances with constrained resources to enable control and to manage over wide area network. They show realization of our proposed architecture by prototyping the system.

In chapter "Towards a Model Level Replication Technique for Fault Tolerant Systems Using AADL", Wafa Gabsi and Bechir Zalila propose a new technique to design replication using the AADL language and its extensibility with property sets. We choose AADL to take advantage of its strong semantics at architecture level.

In chapter "Model Inference of Mobile Applications with Dynamic State Abstraction", Sebastien Salva and Patrice Laurencot and Stassia R. Zafimiharisoa propose an automatic testing method of mobile applications, which also learns formal models expressing navigational paths and application states.

In chapter "Automatic Generation of S-LAM Descriptions from UML/MARTE for the DSE of Massively Parallel Embedded Systems" Manel Ammar, Mouna Baklouti, Maxime Pelcat, Karol Desnos, and Mohamed Abid propose a tool which automates the generation of the System-Level Architecture Model (S-LAM) from a Unified Modeling Language-based (UML) model annotated with the Modeling and Analysis of Real-Time and Embedded Systems (MARTE) profile.

In chapter "Automatic Translation of OCL Meta-Level Constraints into Java Meta-Programs" Sahar Kallel, Chouki Tibermacine, Bastien Tramoni, Christophe Dony and Ahmed Hadj Kacem describe a system that generates metaprograms starting from architecture constraints, written in OCL at the metamodel level, and associated to a specific UML model of an application. These metaprograms enable the checking of these constraints at runtime.

In chapter "Towards a Formal Model for Dynamic Networks Through Refinement and Evolving Graphs" Faten Fakhfakh, Mohamed Tounsi, Ahmed Hadj Kacem and Mohamed Mosbah propose a general and formal model for dynamic networks based on evolving graphs and Event-B formal method. They investigate an example of a distributed algorithm encoded by local computations models.

In chapter "An Iterated Variable Neighborhood Descent Hyperheuristic for the Quadratic Multiple Knapsack Problem", Takwa Tlili, Hiba Yahyaoui, and Saoussen Krichen propose a hyper-heuristic approach based on the iterated variable neighborhood descent algorithm for solving the QMKP. Numerical investigations based on well-known benchmark instances. The results clearly demonstrate the good performance of the proposed algorithm in solving the QMKP.

It is our sincere hope that this volume provides stimulation and inspiration, and that it will be used as a foundation for works to come.

June 2015   Keizo Saisho
                                                                                    Kagawa University, Japan

Contents

Contributors

Noboru Abe Faculty of Information and Communication Engineering, Osaka Electro-Communication University, Neyagawa-shi, Osaka-fu, Japan

Mohamed Abid CES Laboratory, National Engineering School of Sfax, Sfax, Tunisia

Manel Ammar CES Laboratory, National Engineering School of Sfax, Sfax, Tunisia

Koichi Anada Research Institute for Science and Engineering, Waseda University, Tokyo, Japan

Mouna Baklouti CES Laboratory, National Engineering School of Sfax, Sfax, Tunisia

Antoine Bossard Graduate School of Science, Kanagawa University, Hiratsuka, Kanagawa, Japan

Karol Desnos IETR, INSA Rennes, CNRS UMR 6164, UEB, Rennes, France

Christophe Dony Lirmm, Montpellier University, Montpellier, France

Faten Fakhfakh ReDCAD Laboratory, FSEGS, University of Sfax, Sfax, Tunisia

Wafa Gabsi ReDCAD Laboratory, National School of Engineers of Sfax, University of Sfax, Sfax, Tunisia

Teruhisa Hochin Kyoto Institute of Technology, Sakyo-ku, Kyoto, Japan

Taocheng Hu State Key Lab of CAD&CG, Zhejiang University, Hangzhou, China

Kouhei Inoue Faculty of Information and Communication Engineering, Osaka Electro-Communication University, Neyagawa-shi, Osaka-fu, Japan

Ryota Ishibashi Network Technology Laboratories, Nippon Telegraph and Telephone Corp., Tokyo, Japan; Currently with Service Design Division, NTT DOCOMO, Inc., Tokyo, Japan

Ahmed Hadj Kacem ReDCAD Laboratory, FSEGS, University of Sfax, Sfax, Tunisia

Georgios Kalantzis Department of Physics, Florida Atlantic University, Boca Raton, FL, USA

Sahar Kallel Lirmm, Montpellier University, Montpellier, France

Keiichi Kaneko Graduate School of Engineering, Tokyo University of Agriculture and Technology, Koganei, Tokyo, Japan

Taiyou Kikuchi Department of Information Science, Nihon University, Tokyo, Japan

Rhee Man Kil College of Information and Communication Engineering, Sungkyunkwan University, Suwon, Republic of Korea

Kyung Tae Kim College of Information and Communication Engineering, Sungkyunkwan University, Suwon, Republic of Korea

Arata Koike Network Technology Laboratories, Nippon Telegraph and Telephone Corp., Tokyo, Japan

Shinji Koka College of Humanities and Sciences, Nihon University, Tokyo, Japan

Saoussen Krichen LARODEC, Institut Supérieur de Gestion Tunis, Université de Tunis, Tunis, Tunisia

Patrice Laurençot LIMOS CNRS UMR 6158, Blaise Pascal University, Clermont-Ferrand, France

Byungjun Lee College of Information and Communication Engineering, Sungkyunkwan University, Suwon, Republic of Korea

Theodora Leventouri Department of Physics, Florida Atlantic University, Boca Raton, FL, USA

Cheng-Chieh Li Department of Computer Science, Tunghai University, Taichung, Taiwan

Chu-Hsing Lin Department of Computer Science, Tunghai University, Taichung, Taiwan

Jung-Chun Liu Department of Computer Science, Tunghai University, Taichung, Taiwan

Winston Lo Department of Computer Science, Tunghai University, Taichung, Taiwan

Youzou Miyadera Department of Mathematics and Information Science, Tokyo Gakugei University, Tokyo, Japan

Mohamed Mosbah LaBRI Laboratory, University of Bordeaux, Talence, France

Hiroki Nomiya Kyoto Institute of Technology, Sakyo-ku, Kyoto, Japan

Hiroaki Oh Faculty of Information and Communication Engineering, Osaka Electro-Communication University, Neyagawa-shi, Osaka-fu, Japan

Hee Jung Park College of Information and Communication Engineering, Sungkyunkwan University, Suwon, Republic of Korea

Maxime Pelcat IETR, INSA Rennes, CNRS UMR 6164, UEB, Rennes, France

Naoshi Sakamoto Tokyo Denki University, Tokyo, Japan

Sébastien Salva LIMOS CNRS UMR 6158, University of Auvergne, Clermont-Ferrand, France

Charles Shang Lynn Cancer Institute, Boca Raton, FL, USA

Hidenobu Tachibana National Cancer Center Hospital East, Kashiwa, Chiba, Japan

Preecha Tangkraingkij Department of Applied Computer Science, School of Information Technology, Sripatum University, Bangkok, Thailand

Chouki Tibermacine Lirmm, Montpellier University, Montpellier, France

Takwa Tlili LARODEC, Institut Supérieur de Gestion Tunis, Université de Tunis, Tunis, Tunisia

Mohamed Tounsi ReDCAD Laboratory, FSEGS, University of Sfax, Sfax, Tunisia

Bastien Tramoni Lirmm, Montpellier University, Montpellier, France

Hiba Yahyaoui LARODEC, Institut Supérieur de Gestion Tunis, Université de Tunis, Tunis, Tunisia

Takeo Yaku Department of Information Science, Nihon University, Tokyo, Japan

Hee Yong Youn College of Information and Communication Engineering, Sungkyunkwan University, Suwon, Republic of Korea

Jinhui Yu State Key Lab of CAD&CG, Zhejiang University, Hangzhou, China

Stassia R. Zafimiharisoa LIMOS CNRS UMR 6158, University of Auvergne, Clermont-Ferrand, France

Bechir Zalila ReDCAD Laboratory, National School of Engineers of Sfax, University of Sfax, Sfax, Tunisia

On the Accelerated Convergence of Genetic Algorithm Using GPU Parallel Operations

Cheng-Chieh Li, Jung-Chun Liu, Chu-Hsing Lin
and Winston Lo

Abstract The genetic algorithm plays a very important role in many areas of applications. In this research, we propose to accelerate the evolution speed of the genetic algorithm by parallel computing, and optimize parallel genetic algorithms by methods such as the island model. We find that when the amount of population increases, the genetic algorithm tends to converge more rapidly into the global optimal solution; however, it also consumes greater amount of computation resources. To solve this problem, we take advantage of the many cores of GPUs to enhance computation efficiency and develop a parallel genetic algorithm for GPUs. Different from the usual genetic algorithm that uses one thread for computation of each chromosome, the parallel genetic algorithm using GPUs evokes large amount of threads simultaneously and allows the population to scale greatly. The large amount of the next generation population of chromosomes can be divided by a block method; and after independently operating in each block for a few generation, selection and crossover operations of chromosomes can be performed among blocks to greatly accelerate the speed to find the global optimal solution. Also, the travelling salesman problem (TSP) is used as the benchmark for performance comparison of the GPU and CPU; however, we did not perform algebraic optimization for TSP.

Keywords Parallel computing · Genetic algorithm · TSP · GPU computing · Island model · Simulated annealing

C.-C. Li · J.-C. Liu · C.-H. Lin (✉) · W. Lo
Department of Computer Science, Tunghai University, Taichung 40704, Taiwan
e-mail: chlin@thu.edu.tw

C.-C. Li
e-mail: g03350006@thu.edu.tw

J.-C. Liu
e-mail: jcliu@thu.edu.tw

W. Lo
e-mail: winston@thu.edu.tw

© Springer International Publishing Switzerland 2016
R. Lee (ed.), *Software Engineering, Artificial Intelligence, Networking
and Parallel/Distributed Computing 2015*, Studies in Computational Intelligence 612,
DOI 10.1007/978-3-319-23509-7_1

1 Introduction

The genetic algorithm, which imitates the process of natural selection, is a heuristic approach to produce solutions for global search and optimization problems. Basically, it is a high performance, parallel, global search method consisting of five major phases: initialization, evaluation, selection, crossover, and mutation. From continuous iterations of these five phases, an approximately optimal solution can be reached. However, the genetic algorithm itself has some drawbacks, such as premature convergence at local optimal solutions. To solve these issues, this research proposes to combine the simulated annealing method with the genetic algorithm and take advantage of parallel computing with the GPU. For parallel computing, the GPU is fundamentally different from the CPU, due to apparent difference of the number of cores and the computing architecture. Since GPU architecture is based on SIMD, we need to focus on memory access techniques to realize a high level of parallelization for genetic algorithms using GPUs. We also replace the mutation step in genetic algorithm with the simulated annealing method. In this way, the search rate for global optimal solutions of the genetic algorithm is greatly improved. More importantly, GPUs with many thousands of cores can handle a much larger population for genetic algorithms a general purpose CPU. Hence, we are able to use the island model [1–3] to perform block divisions on the large scale population to easily find the global optimal solution for genetic algorithms.

2 Background

2.1 HAS Architecture

Inside conventional computing architecture, only one CPU, or a multi-core CPU handles all operations. To execute massive and high speed operations, lots of CPUs are required, resulting in increasing hardware cost and electric power consumption. The benefit of Heterogeneous System Architecture (HSA) that integrates CPUs and GPUs is that it selects operations of different properties inside an algorithm and sends them to CPUs or GPUs with better hardware architecture for these operations. Thus, we can not only implement optimal hardware architecture for specific algorithms but also execute operations in GPUs and CPUs in parallel to accelerate computations.

2.2 Hardware Architecture of GPU

GPUs originally are hardware designed to handle graphics rendering, but in recent years GPU manufacturing companies, such as AMD and NVIDIA, start developing techniques to utilize the large amount of cores inside GPUs for computation. In this

Fig. 1 SMX architecture

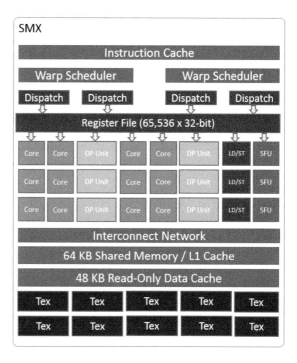

research, we adopt Kepler micro architecture developed by NVIDIA, in which the core unit is called as the Streaming Multiprocessor (SMX). Each SMX consists of 192 CUDA cores. As shown in Fig. 1, each CUDA core can be treated as a thread processing unit. Each GPU chip has more than one SMX core, which explains the advantage of GPU for parallel computing. But since GPUs are limited by adopting Single instruction, multiple data (SIMD) [4, 5] architecture, they are suitable for handling operations of algorithms with data of high independence. For operations of algorithms with data of high dependence amid one another, the performance of GPUs may be lower than that of CPUs.

2.3 Memory Access on GPU

Many ALU cores are provided by GPUs for computation; however, to speed up computation, memory access methods should be designed carefully, since the large amount of cores will produce massive memory requests. When implementing an algorithm using GPU computing, in addition to the possible levels of parallelization for the algorithm, the sequence of memory access and arrangement of data should be carefully considered.

Table 1 Performance of memory access for various settings of cpu and gpu (in ms)

	2047	2048	2049
CPU_host	68.34	115.96	71.0
GPU_read	0.52	0.43	0.44
GPU_write	0.73	0.67	0.66
GPU_r/w	0.40	0.35	0.39

The memory access of GPU follows a special mechanism that if the sequence of data access is consistent with that of thread, then 32 sets of data will be accessed simultaneously in a unit called a warp; if not, the so-called bank-conflict will occur. Table 1 shows performance of CPU and GPU with various settings for data access, in which we observe that GPU_r/w in warp operation is most efficient, even faster than GPU_read or GPU_write.

2.4 Genetic Algorithm

Biological evolution is mainly accomplished through mating between the chromosomes and mutations. The genetic algorithm mimics the biological mechanisms of heredity and evolution. Different problems use different coding methods to represent feasible solutions to the problems, so that a variety of genetic algorithms are constructed with different encoding methods and different genetic operators.

The genetic algorithm in the artificial intelligence is a kind of evolutionary and heuristic algorithms used to search the optimal solutions for problems. Genetic algorithm with well-adjusted parameters can quickly find approximate optimal solutions even for solutions in a very complex space.

Basic steps of the adopted genetic algorithm are listed as follows:

Step 1: randomly generating initial population

Step 2: evaluating each chromosome

Step 3: using a Roulette method to duplicate chromosomes until the amount of population is the same as that of the initial population

Step 4: adjoining chromosomes probabilistically carrying out mating with randomly chosen blocks

Step 5: each chromosome being probabilistically decided whether to execute mutation or not

Step 6: repeating processes from step 1 until a specified stopping criterion is satisfied

2.5 TSP Problem

The travelling salesman problem (TSP) is a well-known mathematical problem. It states that a salesman is going to visit n cities, and he/she needs to select a shortest path to visit each of the n cities once and return to the starting city in the end.

TSP is a very important combinatorial optimization problem. For illustration, we define a problem space in Eq. (1) and an objective function in Eq. (2):

$$X = \{node_0, node_1, node_2, node_3\} \tag{1}$$

$$Minimize\left(f(x) = dist(node_0 x[0]) + \sum_{i=0}^{2} \frac{dist(x[i], x[i+1])+}{dist(x[3], node_0)}\right) \tag{2}$$

Obviously, it is usually not easy to find the minimal solution for above statements. The genetic algorithm can be applied to solve for the approximately optimal solution.

3 Method

3.1 GPU Parallel Computing for Genetic Algorithms

Many parallel algorithms using CPUs have been proposed, but research in parallel algorithm using GPUs is hard to find. Since computation architecture of GPUs adopts SIMD, when implementing algorithms using GPUs, one must take care of data dependence in computing. Thus, it is not easy to transport algorithms running on CPUs to GPUs to speed up computation. For parallel computing of genetic algorithms on GPUs, we examine steps of a genetic algorithm and observe three kinds of problems need to be considered:

1. Random number generation methods in GPUs
2. Highly parallelized duplication process
3. Parallel crossover process

Random number generation plays a very important role for genetic algorithms. It is vital to generate random number in parallel. In the library *curan* provided by NVIDIA offer a random number generation method for the kernel, i.e., every thread inside GPU can separately generate its own random number. Computers can generate random numbers using RNG function. Same random numbers will be generated by a kernel if inputted with the same seed to RNG. The function curandUniform() can be used to shift a state. Before generating next random numbers, the seed of each thread is shifted with various amount, also, states are stored back to the global memory for later computation.

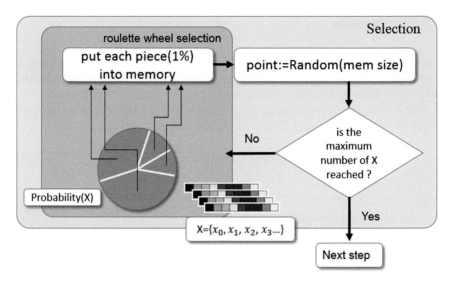

Fig. 2 Selection operation in CPU

During the duplication process of the genetic algorithm, the evaluated chromosomes will be used to generate the next generation population by the Roulette method or other Selection method; chromosomes with higher evaluation have better chance to be chosen. Figure 2 shows the conventional selection algorithms using CPUs.

For the Roulette selection method, Eq. (3) is used:

$$probability(x_i) = \frac{x_i}{\sum_{k=1}^{n} x_k} \times 100\% \tag{3}$$

where x_i is the evaluation value of chromosome i. Every duplicated chromosome needs to go through the probabilistic Roulette operation again, i.e., each chromosome performs one probabilistic operation. Thus, the operation cost using CPU for duplicating chromosome sequence and generating random numbers is of $O(n^2)$; whereas, the operation cost using GPU is of $O(n)$, where the parallel Roulette method is performed by threads for memory access, and then shifted n times to achieve parallel computing. Figure 3 shows the flow for the used Selection operation in GPU. The range of probabilistic outcomes of the traditional Roulette selection method has high dependence on the adopted equations. Also, severe bank-conflicts occur when all threads simultaneously access data. In view of this problem, we perform Selection operation in GPU (see Fig. 3) by Eq. (4):

$$\omega_i = Random(thread_{id}) \bmod \alpha * \frac{1}{x_i} \tag{4}$$

Fig. 3 Selection operation
in GPU

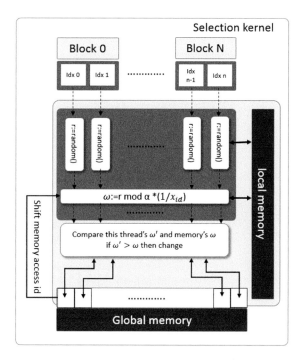

This Selection method uses a shift technique by threads to access memory, and the operation can be done after several shifts of population. To implement Roulette Selection operation, at each shifting operation, the ω value of the accessed memory is replaced by that (i.e., ω') of the thread if $\omega' > \omega$. In this way, not only the thread can access data in memory by a warp per time, but also the huge amount of resource spent on add operations in the traditional Roulette method to generate the probability range is alleviated.

Finally, if every thread is used for one chromosome, parallel access of data cannot be performed for crossover operation that swapping genetic sequences between two chromosomes. So, the number of used threads is halved, and each thread owns different RNG sequence. Figure 4 shows the crossover operation in GPU.

So far, we have described how to perform parallel computing for general genetic algorithms. There is still one unsolved issue for genetic algorithms, i.e., premature convergence at local optimal solutions. To deal with this problem, we use a parallel simulated annealing method described in the following subsection.

3.2 Parallel Simulated Annealing Method

The simulated annealing method, a generic probabilistic metaheuristic for the global optimization problem, is inspired by annealing in metallurgy. After heating and

Fig. 4 Crossover operation in GPU

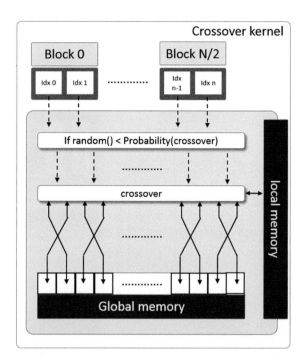

controlled cooling of a material such as iron, its grains will leave original place and move randomly to other locations, thus, reducing defects in the crystal lattice.

For the simulated annealing method, two parameters representing energy and the annealing rate are manually set, where energy gives tolerance of levels of difference and the annealing rate indicates the rate of decline in energy. With lower energy, it is more difficult to leave local minimums. By combining simulated annealing with the genetic algorithm, the mutation process of the genetic algorithm is replaced with the annealing method instead. In this way, it has a higher probability to leave local optimal solutions. The algorithm of the simulated annealing is shown in Table 2.

Table 2 Algorithm of simulated annealing, where acceptance probability function p() is given in Eq. (5)

S := s0;

e := E(s);

while k < kmax && e > emax

sn := neighbor (s)

de := E(sn)

if random() < P(e, de, temp (k/kmax)) then

s:= sn

e:= de

k :=k+1

return s

Table 3 Pseudo codes of genetic algorithm combined with simulated annealing

---

E:=evaluation(x);

d_E:=evaluation(Mutation(x));

delta:=d_E − E;

if delta>0 and exp(-delta/ Temperature) then

E:=d_E;

$$P = \begin{cases} 1 & if\ \Delta E \leq 0 \\ e^{\frac{-\Delta E}{T}} & if\ \Delta > 0 \end{cases} \tag{5}$$

According to the result of acceptance probability function P(), we decide whether to move the grain or not according to the amount of changes of energy.

To apply simulated annealing to genetic algorithm, in Table 2 variable S is treated as the state of a chromosome; e, the current evaluation value; de, the evaluation value after mutation. When the outcome of function P() is larger than that of the random number, this chromosome succeeds in mutation. Table 3 lists the pseudo codes of the genetic algorithm combined with the simulated annealing method.

As Temperature decreases, the allowable delta value for chromosomes to perform mutation becomes smaller; and at some point when Temperature is close to 0, chromosomes will stop mutation.

The proposed mutation method has better random search ability than traditional mutations with a fixed mutation rate. However, if Temperature decreases too fast, the genetic algorithm might converge to local optimal solutions; on the other hand, if Temperature decreases too slowly, the efficiency of the genetic algorithm is affected. To prevent this premature convergence issue, a polynomial function is used, as shown in Eq. (6):

$$Temperature := T_0 * ROD_0^x + T_1 * ROD_1^x...T_n * ROD_n^x \tag{6}$$

where x represents the iteration number; both ROD_i and T_i are smaller than 1, and ROD_i increases (and T_i decreases) as subscript i increases from 0 to n.

3.3 Modified Parallel Genetic Algorithm

As described at above subsections, computations of all phases of genetic algorithms have been parallelized to accelerate iteration, i.e., evolution speed. Besides, to solve premature convergence problem, the simulated annealing method has been applied. However, since the simulated annealing method itself probabilistically converges to global optimal solutions, it may fall into local optimal solutions by chance. As inspired by the Island model, we use the concept of the block method to divide

Fig. 5 Optimized parallel
genetic algorithm using the
Island model

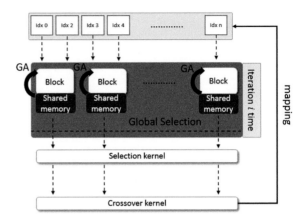

the whole population into several blocks. Each chromosome is still computed by a
thread. By dividing into blocks, interference among individual blocks is prevented:
each block is provided with a different initial population, and will converge to a
local or global optimal solution. After performing several iterations of the genetic
algorithm at each block, a global selection operation is conducted. In this way, we
have better chance to find the global optimum. Figure 5 shows the flow chart of the
optimized parallel genetic algorithm using the Island model [1, 2, 6].

4 Result and Discussion

The setting of experiments is as follows.

For the CPU, we used Intel(R) Core(TM)i7 CPU 4770K @3.5GHz and memory
of 8.00 GB; for the GPU, we used Nvidia Titan Black. The used operating system
is Linux Ubuntu 14.04(64bit). The used CPU codes to solve TSP using genetic
algorithms references the following website: http://simulations.narod.ru/.

We did not optimize the TSP algorithm [7] in this study, since we focus on developing optimization for genetic algorithms and only use TSP as the benchmark for
performance comparison using the GPU or CPU.

4.1 Population versus Iteration Speed

To compare performance of genetic algorithms of CPUs with that of GPUs, we first
performed 2000 iterations of the genetic algorithm with various populations and
nodes. The time costs of the CPU in seconds are listed in Table 4 and illustrated in

Table 4 CPU computation time (in seconds) of various populations and nodes for 2000 iterations

		Nodes				
		16	32	64	128	256
Population	512	20	25.95	63.25	120.15	251.32
	1024	28.01	46.43	115.39	232.43	481.84
	2048	47.03	101.46	224.33	454.25	1000
	4096	109.62	233.87	438.87	892	1969.5
	8192	270.75	465.64	885.5	1807.75	3964

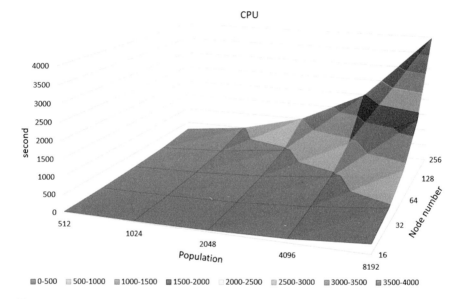

Fig. 6 CPU computation time (in seconds) of various populations and nodes for 2000 iterations

Fig. 6. As shown in Fig. 6, the computing time of the CPU increases very fast as the population or the number of nodes increases. The time costs of the GPU in seconds are listed in Table 5 and illustrated in Fig. 7. As shown in Fig. 7, the computing time of the CPU increases slowly as the population or the number of nodes increases. The speedups of GPU versus CPU are listed in Table 6 and illustrated in Fig. 8. As shown in Fig. 8, significant speedups (more than two orders of magnitude) are obtained when the population or the number of nodes increases.

Table 5 GPU computation time (in seconds) of various populations and nodes for 2000 iterations

		Nodes				
		16	32	64	128	256
Population	512	0.445	0.758	1.245	2.171	5.202
	1024	0.465	0.789	1.299	2.265	5.355
	2048	0.502	0.857	1.412	2.574	6.488
	4096	0.557	1.017	1.731	3.401	8.395
	8192	0.734	1.326	2.694	7.058	16.399

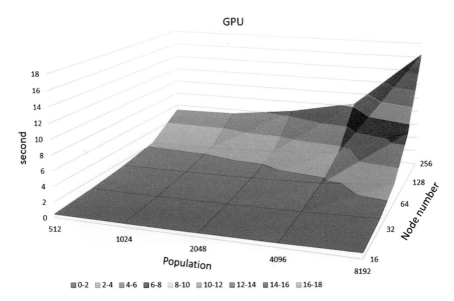

Fig. 7 GPU computation time (in seconds) of various populations and nodes for 2000 iterations

4.2 Converging Rates with Various Node Numbers

In this experiment, we investigate the relative errors or deviations from the global optimal solution of the CPU and GPU. We use Eq. (7) to calculate the relative error δ:

$$\delta = \left| \frac{v - v_{approx}}{v} \right| \times 100\,\% \tag{7}$$

Figures 9 and 10 show the relative errors over 5 min of computation using the CPU and GPU, respectively. We observe that when the number of nodes increases, the relative errors increase as expected, as it becomes harder to converge. Also, the relative errors for initial populations are usually very high.

Table 6 Speedup of computation time for gpu versus cpu

		Nodes				
		16	32	64	128	256
Population	512	44.94	34.23	50.8	55.34	48.31
	1024	60.23	58.84	88.82	102.61	89.97
	2048	93.68	118.38	158.87	176.47	154.13
	4096	196.8	229.96	253.53	262.27	234.6
	8192	368.86	351.16	328.69	256.12	241.72

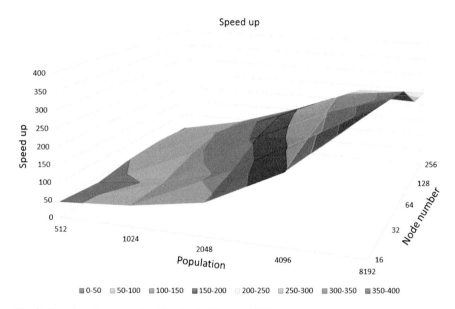

Fig. 8 Speedup of computation time for GPU versus CPU

Figures 11 and 12 show the relative errors over 5000 iterations of computation using the CPU and GPU, respectively. From these results, we observe that compared with the CPU the optimized GPU parallel genetic algorithm also has better convergence with the same number of iterations.

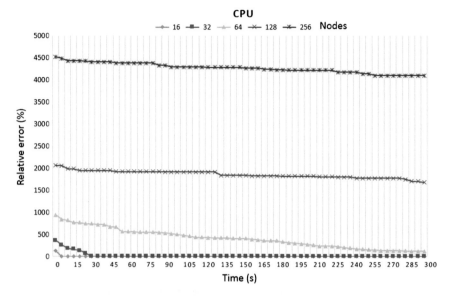

Fig. 9 Relative errors (in percentage) using CPU in five minutes of computation

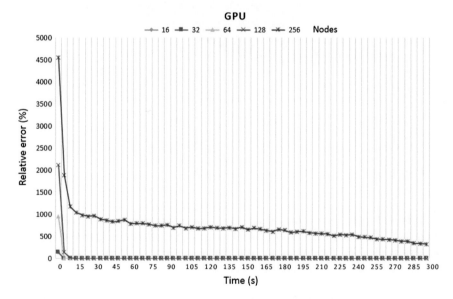

Fig. 10 Relative errors (in percentage) using GPU in five minutes of computation

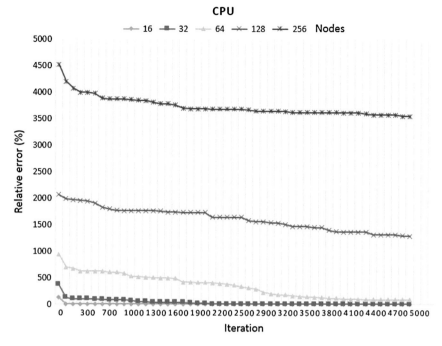

Fig. 11 Relative errors (in percentage) using CPU over 5000 iterations of computation

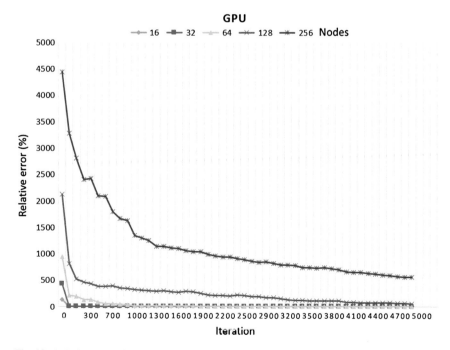

Fig. 12 Relative errors (in percentage) using GPU over 5000 iterations of computation

5 Conclusion

In this study, we use GPUs instead of CPU to accelerate computations of parallel genetic algorithms. To optimize genetic algorithms, a simulated annealing method is used for mutation operations; however, simulated annealing itself has shortcomings that temperature and the temperature decreasing rate should be manually adjusted. Thus, we modify the simulated annealing method with a polynomial Temperature function to attain efficient mutation operations. As inspired by the Island model, we also partition population into several blocks (islands) of threads (individuals) to have better chance to find the global optimum. The simulation results show that when population or the number of nodes increase, performance of optimized parallel genetic algorithms using GPUs is much superior to that using CPUs, in regard with both the convergence speed and computation time cost. We conclude that by stripping out most of the operations with high dependence, the genetic algorithm can be modified to achieve high performance parallel computing on GPUs with SIMD architecture.

Acknowledgments This research was partly supported by Ministry of Science and Technology, Taiwan, under grant number MOST 103-2221-E-029 -020.

References

1. Whitley, D., Rana, S., Heckendorn, R.B.: The Island model genetic algorithm: on separability, population size and convergence. J. Comput. Inf. Technol. **7**, 33–47 (1999)
2. Darrell, W., Rana, S., Heckendorn, R.B.: Island model genetic algorithms and linearly separable problems. Evolutionary Computing, pp. 109–125. Springer, Berlin (1997)
3. Scott Gordon, V., Darrell Whitley, L.: Serial and parallel genetic algorithms as function optimizers. In: Proceedings of the 5th International Conference on Genetic Algorithms, pp. 177–183 (1993)
4. Garland, M., Grand, S.L., Nickolls, J., Anderson, J., Hardwick, J., Morton, S., Phillips, E., Zhang, Y., Volkov, V.: Parallel computing experiences with CUDA. IEEE Micro **28**, 13–27 (2008)
5. Nickolls, J., Buck, I., Skadron, K., Garland, M.: Scalable parallel programming with CUDA. ACM Queue **6**(2), 40–53 (2008)
6. Melab, N., Talbi, E.-G.: GPU-based island model for evolutionary algorithms. In: Proceedings of the 12th Annual Conference on Genetic and Evolutionary Computation, New York, USA, pp. 1089–1096. ACM Press (2010)
7. Grefenstette, J.J., Gopal, R., Rosmaita, B., Van Gucht, D.: Genetic algorithm for the traveling salesman problem. In: Proceedings of International Conference on Genetic Algorithms and their Applications, pp. 160–165 (1985)

A GPU-Based Pencil Beam Algorithm for Dose Calculations in Proton Radiation Therapy

Georgios Kalantzis, Theodora Leventouri, Hidenobu Tachibana and Charles Shang

Abstract Recent developments in radiation therapy have been focused on applications of charged particles, especially—protons. Proton therapy can allow higher dose conformality compared to conventional radiation therapy. Dose calculations have an integral role in the successful application of proton therapy. Over the years several dose calculation methods have been proposed in proton therapy. A common characteristic of all these methods is their extensive computational burden. One way to ameliorate that issue is the parallelization of the algorithm. Graphics processing units (GPUs) have recently been employed to accelerate the proton dose calculation process. Pencil-beam dose calculation algorithms for proton therapy have been widely utilized in clinical routine for treatment planning purposes in most clinical settings, due to their simplicity of calculation scheme and acceptable accuracy. In the current study a GPU-based pencil beam algorithm for dose calculations with protons is proposed. The studies indicated a maximum speedup factor of ~ 127 in a homogeneous phantom.

Keywords Pencil-beam · Proton therapy · GPU · Dose calculations

1 Introduction

Radiation therapy uses high-energy radiation to shrink tumors and kill cancer cells by damaging their DNA. Contemporary radiation therapy often involves the process where radiation beams are delivered to the cancer site from external beams. In a clinical setup, external radiation therapy most commonly involves high energy photons

G. Kalantzis (✉) · T. Leventouri
Department of Physics, Florida Atlantic University, Boca Raton, FL 33431, USA
e-mail: gkalan@gmail.com

H. Tachibana
National Cancer Center Hospital East, Kashiwa, Chiba 277-8577, Japan

C. Shang
Lynn Cancer Institute, Boca Raton, FL 33486, USA

© Springer International Publishing Switzerland 2016
R. Lee (ed.), *Software Engineering, Artificial Intelligence, Networking and Parallel/Distributed Computing 2015*, Studies in Computational Intelligence 612,
DOI 10.1007/978-3-319-23509-7_2

or electrons. However, in 1946 Wilson demonstrated the potential clinical benefits of protons for cancer treatment [1] and the first clinical evaluations of external radiation therapy with protons appeared by Lawrence et al. [2]. The dosimetric properties that establish protons attractive for radiation therapy is their nearly straight line trajectories in the matter and a narrow Bragg peak which results in insignificant dose downstream the treatment sites. That implies increased local tumor control while sparing normal tissue [3].

Dose calculations play a crucial role in proton therapy treatment. In general, the gold standard for radiotherapy dose calculations is Monte Carlo (MC) methods. However, MC simulations for charged particles is a time-consuming and computational demanding task. In the time-critical clinical environment, fast dose calculations ensure a smooth workflow, and also offer planners the opportunity to fine tune the treatment planning for each individual patient. The necessity of fast dose calculations becomes more apparent in intensity-modulated proton therapy [4, 5], and 4D treatment planning [6, 7]. One way to ameliorate that issue is to parallelize the computational method.

Recent developments, have demonstrated the potential of Graphics processing Units (GPUs) for MC simulations for proton therapy. In particular, Kohno et al. [8] developed a simplified MC method (SMC) for proton transportation on a GPU platform. The SMC was compatible with GPU's SIMD structure in the sense that each GPU thread calculated the transportation of the proton while all of the threads performed the same instructions on different data depending on the current proton status. Track-repeating methods have also been utilized in the past for fast GPU-based MC simulations [9, 10]. In these methods, a database of proton transport histories is generated in advance for a homogeneous water phantom using an accurate MC code. Then, for a patient case, the track-repeating MC calculates dose distributions by repeating proton tracks from the database and scaling the step length and scattering angle within each track according to the tissue densities.

An alternative method is the pencil beam algorithms (PBAs) for dose calculations in radiation therapy. This method has been widely utilized in daily clinical applications due to its simplicity of calculation scheme and acceptable accuracy in many clinical cases. In PBAs a broad beam is divided into a number of small rectangular (beamlets), and the dose contribution of each beamlet to every voxel is based on analytical or empirical calculations. PBAs are well suited for parallelization on GPUs since each thread can calculate the dose for each beamlet. Gu et al. [11] have reported 200–400 speedup factors for a GPU-based pencil beam method for intensity modulation radiation therapy, while Fujimoto el al. [12] reported speedups up to 20 for proton calculations.

Foca, a Matlab-based treatment planning software, was recently presented for proton radiotherapy [13]. The aforementioned software was designed primarily for research and educational purposes. In Foca, dose calculations for both, active (pencil-beam scanning) and passive (double scattering) modalities have been implemented. The main advantages of Matlab are its intuitive higher-level syntax, advanced visualization capabilities and the availability of toolboxes with several numerical methods.

In the current study we present for the first time, to our best knowledge, a GPU-based PBA for proton dose calculations in Matlab. The evaluation of our method was established on the speedup factors for square radiation fields in homogenous phantom of water.

2 Pencil-Beam Algorithm for Protons

2.1 Depth Dose Distributions

In the current study we employed an analytical expression for the protons depth dose distribution [14]. For a monoenergetic proton beam along the z axis, impinging on a homogeneous medium at z = 0 the energy fluence, Ψ, at depth z in the medium can be written in the form:

$$\Psi(z) = \Phi(z) \cdot E(z) \tag{1}$$

$\Phi(z)$ is the particle fluence, i.e. the number of protons per cm^2, and E(z) is the remaining energy at depth z. The total dose, D(z) is given by:

$$D(z) = -\frac{1}{\rho}\left(\Phi(z)\frac{dE(z)}{dz} + \gamma\frac{d\Phi(z)}{dz}E(z)\right) \tag{2}$$

where, γ is the fraction of the energy released in the inelastic nuclear interactions and absorbed locally, and ρ is the mass density of the medium. In order to determine this depth-dose curve, we only need to know the functions E(z) and $\Phi(z)$. The relationship between the initial energy E(z = 0) = E_0 and the range z = R_0 in the medium is approximately given by:

$$R_0 = aE_0^p \tag{3}$$

With p = 1.5, this relationship is known as Geiger's rule, which is valid for protons with energies up to about 10 MeV. For higher energies used in radiation therapy the exponent p increases to 1.8. The factor α is approximately proportional to the square root of the effective atomic mass of the absorbing medium. It should be noted that this simplified range-energy relationship is valid in good approximation for arbitrary media and various particles with atomic masses approximately between one (protons) and twelve (carbon ions) [15]. The beam deposits energy in the medium along its path from z = 0 to z = R_0. Thus, according to the range-energy relationship:

$$E(z) = \frac{1}{a^{1/p}}(R_0 - z)^{1/p} \tag{4}$$

Lee et al. [16] have derived the proportionality for the protons fluence:

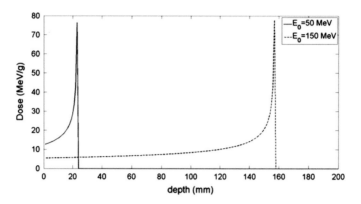

Fig. 1 Depth-dose distribution for 50 MeV (*solid line*) and 150 MeV (*dashed line*) protons in water

$$\Phi(z) \propto \frac{1}{1 - P(R_0 - z)} \tag{5}$$

Where P is the probability that protons may be lost from the beam due to nuclear interactions. By normalizing to the initial fluence Φ_0 we obtain:

$$\Phi(z) = \Phi_0 \frac{1 + \beta(R_0 - z)}{1 + \beta R_0} \tag{6}$$

Finally, the depth-dose distribution can be calculated as follows for $z < R_0$:

$$D(z) = \Phi_0 \frac{(R_0 - z)^{1/p-1} + (\beta + \gamma\beta p)(R_0 - z)^{1/p}}{\rho p a^{1/p}(1 + \beta R_0)} \tag{7}$$

where the slope parameter β was determined to be $\beta = 0.012 \, \text{cm}^{-1}$. Expression (7) gives D in units of MeV/g, if ρ is given in g/cm^3. To obtain D in Gy, one needs to multiply by the factor $10^9 \text{e/C} = 1.602 \times 10^{-10}$, where e is the elementary charge.

Figure 1 illustrates the depth-dose distributions in water for protons with initial energy 50 MeV (solid line) and 150 MeV (dashed line) respectively. Similarly, Fig. 2 shows the proton energy as a function of depth in water for initial energy 50 MeV (solid line) and 150 MeV (dashed line) respectively.

2.2 Spread-Out Bragg Peak

It is well known that a pristine Bragg peak is not wide enough to cover most treatment volumes and therefore unsuitable for cancer treatment. Rather, it is necessary to "spread out" the Bragg peak to deliver uniform dose within the target volume, by

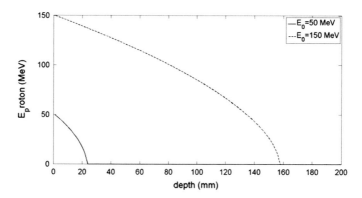

Fig. 2 Protons energy as a function of depth in water for 50 MeV (*solid line*) and 150 MeV (*dashed line*) E_0

providing a suitably weighted energy distribution of the incident beam. For that purpose poly-energetic pencil beams are utilized, which are individually adapted to the proximal and distal edge of the target volume, such that the dose is constant along the depth of the target volume. This creates a highly conformal high dose region, e.g., created by a spread-out Bragg peak (SOBP) with the possibility of covering the tumor volume with high accuracy. At the same time this technique delivers lower doses to healthy tissue than conventional photon or electron techniques. The mathematical problem consists of the calculation of weighting factors $W(R)$ for the Bragg peaks such that the superposition results in a flat SOBP of height D_0 within an interval $[d_a, d_b]$. Bortfeld et al. [17] have derived an analytical formulation of the weights $W(R)$ as follows:

$$W(R) = \begin{cases} \rho D_0 \frac{p\sin(\pi/p)a^{1/p}}{\pi(d_b-R_0)^{1/p}}, & d_a \leq R_0 < d_b \\ 0 & , R_0 < d_a, R_0 > d_b \end{cases} \tag{8}$$

The shape of the SOBP curve can now be calculated by estimating the $D(z)$ (Eq. 7) and $W(R)$. In the special case of $p = 1.5$ and $r < 10$, the depth dose of the SOBP, $D_{SOBP}(r)$ can be approximated by a simpler relationship:

$$D_{SOBP}(r) \approx \frac{D_0}{1 + 0.44r^{0.6}} \tag{9}$$

Deviations of Eq. 9 are within $\pm 1.5\%$ of D_0. Figures 3 and 4 illustrate a set of weighted Bragg peaks and their superposition creating a SOBP respectively.

Fig. 3 A set of weighted pristine Bragg peaks of different energies

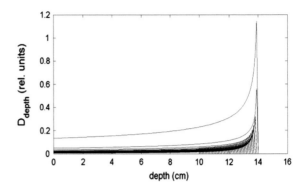

Fig. 4 Superposition of weighted Bragg peaks creating a SOBP

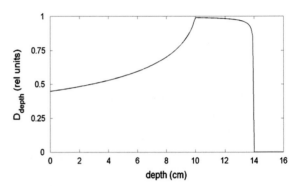

2.3 Lateral Profiles of Pencil Beams

Protons are assumed to emanate axial symmetrically from the finite-sized source with a specified mean residual range and

with no angular emittance initially. For simplicity we assumed scattering only in the patient, which is increasing the radial emittance at the depth of the point of interest. The dose at a point of interest is then determined, by the residual range of protons directed at that point. This is equal to the residual range of protons entering the patient, minus the radiographic path length from the surface to the point of interest, $r(x_p, y_p, z_p)$. The radiographic path length from the surface to the point of interest, rpl_p is based on a pixel-by-pixel integration through a CT study, given by:

$$rpl_p = \int_{surface}^{z_p} dz' WED(CT(z')) \tag{10}$$

where $CT(z')$ is the CT value at the point of distance z' along the path of integration, and WED is value from a look-up Table which converts between CT value and water-equivalent density, as described by Chen et al. [18]. To compute the dose at a point of interest from a given pencil beam, we follow the approach of Hogstrom et al. [19]

who reported on a pencil beam algorithm to model electron beams. The dose, $d(x', y', z')$, at a point (x', y', z'), due to pencil beam is separated into a central-axis term, $C(z')$, and an off-axis term, $O(x', y', z')$:

$$d(x', y', z') = C(z') \cdot O(x', y', z') \tag{11}$$

The central-axis term is taken from the broad-beam central-axis depth dose in water modified by an inverse square correction:

$$C(z') = D(d_{eff}) \cdot \left(\frac{ssd + d_{eff}}{z'} \right)^2 \tag{12}$$

where $D(d_{eff})$ is described in Eq. 7, ssd is the source-to-surface distance and d_{eff} is the effective depth given by:

$$d_{eff} = R_0 - (R_r - rpl(z')) \tag{13}$$

The off-axis term $O(z')$ is taken to be the lateral flux distribution from the radial emittance suffered by protons directed along the axis of the pencil beam. The distribution is considered Gaussian [20]:

$$O(x', y', z') = \frac{\exp(-(x'^2 + y'^2)/2\sigma(z')^2)}{2\pi\sigma(z')^2} \tag{14}$$

where:

$$\frac{\sigma(z')}{\sigma(R_0)} = a_1 \frac{z'}{R_0} + a_2 \frac{z^2}{R_0^2} \tag{15}$$

and,

$$\sigma(R_0) = 0.02275R_0 + 0.12085 \cdot 10^{-4} \cdot R_0^2 \tag{16}$$

Figure 5 depicts plots of Eq. 14 for 150 MeV monoenergetic protons beam for various depths. We notice the increased variance, which reflects to larger penumbra of the PB, as the depth increases.

2.4 GPU-Based Parallelization

The pencil beam method can be used to represent an incident broad beam as a collection of infinitesimally narrow pencil beams (PBs). The major assumptions in a PB dose calculation algorithm include: (1) the broad beam from a point source can be divided into identical beamlets, and (2) the total dose to a point P(x, y, z) is equivalent to integration over pencil beam dose contributions to P(x, y, z). A total dose distribution is obtained by a superposition of the contributions from elementary

Fig. 5 Lateral distributions as a function of distance from the central axis for various depths d in cm

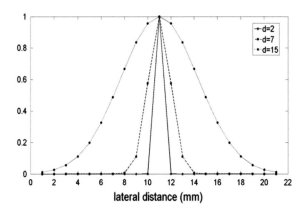

Fig. 6 Schematic representation of dose deposition into a particular voxel (x, y, z) from the PB$_i$. The lateral Gaussian-shaped curves represent the lateral penumbra of the PB$_i$

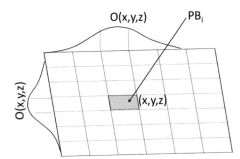

pencil beams. The total dose D(x, y, z) at point (x, y, z) is calculated as follows:

$$D(x, y, z) = \iint dx'dy' C^{(x', y')}(z) O(x - x', y - y', z) \qquad (17)$$

$C^{(x', y')}(z)$ indicates the central-axis term of the beam passing through a point (x', y', z). The second term, $O(x-x', y-y', z)$, is obtained from Eq. 14. In practice, the integration of Eq. 17 is discretized. A region of interest is divided into small elements, which are called voxels. The voxel size is 1 or 2 mm in general. The discretization of equation (17) leads to the following expression for the dose at voxel (i, j, k):

$$D(i, j, k) = \sum_{i', j'} C^{(i', j')}(k) w(i', j', i, j) \qquad (18)$$

where, w(i', j', i, j) indicates the dose contribution to a voxel (i, j) from the beam passing through a voxel (i', j') within the k plane. Figure 6 illustrates a graphical representation of the PB dose deposition method.

For the parallelization of the dose calculations, a native? approach would be that each thread in the GPU to handles a single PB and calculates the dose at adjacent points according to Eq. 17. A different method would be, that each GPU in each thread concentrates on one particular calculation point (or else voxel) and accumulates the dose deposited by every single PB. The main disadvantage of the former method is the atomic operators, which are required in the case when more than one thread tries to write the deposited dose at the same voxel. In addition, for our case the code was implemented in MATLAB and we do not have access to each individual thread, rather algebraic operations of matrices are parallelized internally by the software by utilizing the high performance library CUBLAS. Therefore, we followed a different approach for the parallelization scheme. The dose calculations for the whole volume of each particular PB was parallelized on the GPU. That can be described by the following three steps:

(1) Calculate C(z) according to Eq. 12 along each PB;
(2) Calculate O(x, y, z) according to Eq. 14 along each beam path;
(3) Calculate the dose at each voxel according to Eq. 17.

Once the dose was calculated for a PB, the summation of the newly calculated dose at each voxel was also established on the GPU and the whole process is repeated for all the PBs.

3 Results for Homogeneous Phantom

3.1 Materials and Elements

In this section we introduce our computation environment and report our results of our method. The serial code was implemented in MATLAB and was launched on a desktop with a quad core Intel Xeon X5550 at 2.67 GHz with 6 GB of RAM. For the parallelization on the GPU the parallel computing toolbox was employed and the code was launched on a GTX 770 with Kepler architecture, 1,536 CUDA cores grouped into eight SMs operating at 1046 MHz and 2 GB of GDRR5 global memory. The performance comparison was established on the speedup factors as described by Eq. 19.

$$speedup = \frac{T_{CPU}}{T_{GPU}} \qquad (19)$$

3.2 Dose Distribution in Water Phantom

Figure 7 illustrates a 2D dose distribution of a 1 mm^2 proton PB of 150 MeV in water phantom. We notice the increased dose at the end of the particle trajectory due to the Bragg peak, as well as, the elevation of the off-axis dose as the depth increases. Isodose line distributions are shown in Fig. 8 for a 25 × 25 mm^2 proton beam of 50 MeV. The x-axis represents the depth of the beam in the water phantom, and the

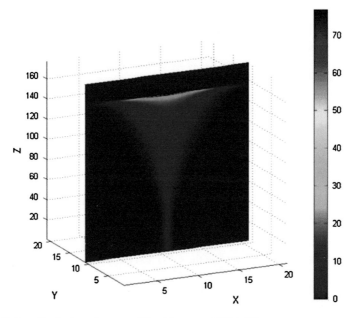

Fig. 7 2D dose distribution of a proton PB with energy 150 MeV in water

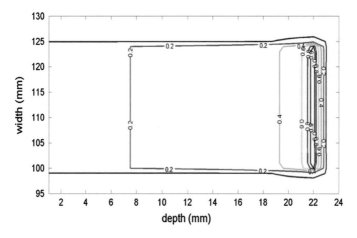

Fig. 8 Isodose distribution for a $25 \times 25\,\mathrm{mm}^2$ proton beam of 50 MeV in water

y-axis is the width of the open beam. Similarly we may observer the lateral spread of the beam as the depth increases. Finally, Fig. 9 depicts lateral profiles for a proton beam of 50 MeV for four different depths. It is worth mentioning, that the units of y-axis are arbitrary units of the normalized dose to the max.

Fig. 9 Lateral profiles for a $25 \times 25\,\text{mm}^2$ proton beam of 50 MeV in water at depth 2, 10, 15 and 20 mm

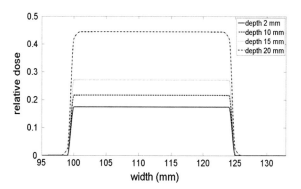

3.3 Evaluation of the Algorithm Performance

The performance of the GPU code was evaluated for three different energies: low (50 MeV), medium (100 MeV) and high (150 MeV). Four square fields were selected for each energy, and the dose calculations were performed with both the serial and parallel codes for a homogeneous water phantom with size $300 \times 300 \times 300\,\text{mm}^3$. The resolution of the PBs was set to 1.0 mm. Table 1 reports the details of the simulations we used in the current study.

Figure 10 illustrates the speedup factors obtained from our simulation results. We notice that the speedup is decreased for smaller number of PBs and smaller energies. We speculate that this is due to the communication overhead between the GPU and the CPU during the iterations of the dose calculations for each PB. The maximum speedup of $\sim$127 was achieved for the highest energy and the largest field size.

Table 1 Geometries used in performance evaluation

E (MeV)	PBs	T_{CPU} (sec)	T_{GPU} (sec)
50	25×25	50.8	1.1
50	50×50	199.3	3.4
50	80×80	515.6	8.5
50	100×100	800.1	13.3
100	25×25	115.0	1.4
100	50×50	461.9	4.8
100	80×80	1.16×10^3	11.6
100	100×100	1.82×10^3	18.3
150	25×25	208.4	1.8
150	50×50	827.5	6.7
150	80×80	2.1×10^3	16.7
150	100×100	3.3×10^3	25.9

Fig. 10 Speedup factors as a
function of energy and
number of PBs

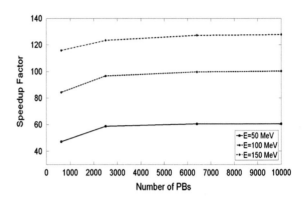

4 Conclusions

A GPU-based PB algorithm for proton dose calculations in Matlab was presented.
A maximum speedup of $\sim$127 was achieved. Future directions of the current work
include extension of our method for dose calculation in heterogeneous phantoms.
Work along this line is in progress.

References

1. Wilson, R.R.: Radiological use of fast protons. Radiology **47**, 487–91 (1946)
2. Lawrence, J.H., Tobias, C.A., Born, J.L., Mc, C.R., Roberts, J.E., Anger, H.O., Low-Beer, B.V., Huggins, C.B.: Pituitary irradiation with high-energy proton beams: a preliminary report. Cancer Res. **18**, 121–34 (1958)
3. Koehler, A.M., Preston, W.M.: Protons in radiation therapy. Comparative dose distributions for protons, photons, and electrons. Radiology **104**, 191–195 (1972)
4. Lomax, A.J., Boehringer, T., Coray, A., Egger, E., Goitein, G., Grossmann, M., Juelke, P., Lin, S., Pedroni, E., Rohrer, B., Roser, W., Rossi, B., Siegenthaler, B., Stadelmann, O., Stauble, H., Vetter, C., Wisser, L.: Intensity modulated proton therapy: a clinical example. Med. Phys. **28**, 317–324 (2001)
5. Oelfke, U., Bortfeld, T.: Inverse planning for photon and proton beams. Med. Dosim. **26**, 113–124 (2001)
6. Paganetti, H., Jiang, H., Trofimov, A.: 4D Monte Carlo simulation of proton beam scanning: modelling of variations in time and space to study the interplay between scanning pattern and time-dependent patient geometry. Phys. Med. Biol. **50**, 983–990 (2005)
7. Kang, Y., Zhang, X., Chang, J.Y., Wang, H., Wei, X., Liao, Z., Komaki, R., Cox, J.D., Balter, P.A., Liu, H., Zhu, X.R., Mohan, R., Dong, L.: 4D proton treatment planning strategy for mobile lung tumors. Int. J. Radiat. Oncol. Biol. Phys. **67**, 906–914 (2007)
8. Kohno, R., Hotta, K., Nishioka, S., Matsubara, K., Tansho, R., Suzuki, T.: Clinical implementation of a GPU-based simplified Monte Carlo method for a treatment planning system of proton beam therapy. Phys. Med. Biol. **56**, N287–294 (2011)
9. Yepes, P.P., Mirkovic, D., Taddei, P.J.: A GPU implementation of a track-repeating algorithm for proton radiotherapy dose calculations. Phys. Med. Biol. **55**, 7107–7120 (2010)

10. Yepes, P.P., Brannan, T., Huang, J., Mirkovic, D., Newhauser, W.D., Taddei, P.J., Titt, U.: Application of a fast proton dose calculation algorithm to a thorax geometry. Radiat. Meas. **45**, 1367–1368 (2010)
11. Gu, X., Choi, D., Men, C., Pan, H., Majumdar, A., Jiang, S.B.: GPU-based ultra-fast dose calculation using a finite size pencil beam model. Phys. Med. Biol. **54**, 6287–6297 (2009)
12. Fujimoto, R., Kurihara, T., Nagamine, Y.: GPU-based fast pencil beam algorithm for proton therapy. Phys. Med. Biol. **56**, 1319–1328 (2011)
13. Sanchez-Parcerisa, D., Kondrla, M., Shaindlin, A., Carabe, A.: FoCa: a modular treatment planning system for proton radiotherapy with research and educational purposes. Phys. Med. Biol. **59**, 7341–7460 (2014)
14. Bortfeld, T.: An analytical approximation of the Bragg curve for therapeutic proton beams. Med. Phys. **24**, 2024–2033 (1997)
15. Raju, M.R.: Heavy Particle Radiotherapy (1980)
16. Lee, M., Nahum, A., Webb, S.: An empirical method to build up a model of proton dose distribution for a radiotherapy treatment planning package. Phys. Med. Biol. **38**, 989–998 (1993)
17. Bortfeld, T., Schlegel, W.: An analytical approximation of depth-dose distributions for therapeutic proton beams. Phys. Med. Biol. **41**, 1331–1339 (1996)
18. Chen, G.T., Singh, R.P., Castro, J.R., Lyman, J.T., Quivey, J.M.: Treatment planning for heavy ion radiotherapy. Int. J. Radiat. Oncol. Biol. Phys. **5**, 1809–1819 (1979)
19. Hogstrom, K.R., Mills, M.D., Almond, P.R.: Electron beam dose calculations. Phys. Med. Biol. **26**, 445–459 (1981)
20. Hong, L., Goitein, M., Bucciolini, M., Comiskey, R., Gottschalk, B., Rosenthal, S., Serago, C., Urie, M.: A pencil beam algorithm for proton dose calculations. Phys. Med. Biol. **41**, 1305–1330 (1996)

Incremental Max-Margin Learning for Semi-Supervised Multi-Class Problem

Taocheng Hu and Jinhui Yu

Abstract In this paper, we proposed an incremental max-margin model for semi-supervised multi-classification learning, where efficient and accuracy need to be considered. Three notable properties are introduced: (1) the model predicts a label for unlabeled sample instance in runtime, and trained with the complete sample instance, while unlabeled and labeled sample instances are unified in our objective function; (2) since the objective function of our model is convex, we can design efficient online algorithm with logarithmic regret, it achieve accurate solution with very little overhead; (3) our model is max-margin machine, which provide our model with considerable generalization capability for future unseen data. Our approach captures essences of the exploration-exploitation tradeoff.

Keywords Max-margin learning · Online algorithm · Multi-classification · Semi-supervised learning

1 Introduction

Given a set of training instance $(\mathbf{x}_t, y_t) \in \mathcal{X} \times \mathcal{Y}, t = 1, \cdots, T$ from a sample space $\mathcal{X}$ and label space $\mathcal{Y}$, classification learning algorithm try to find a classifier h from a domain $\mathcal{X}$ to a label space $\mathcal{Y}$, and the performance of a prediction is measured by the probability if $h(\mathbf{x})$ is the correct label. It is a basic problem in machine learning, surfacing a variety of domains, including object recognition, speech recognition, document categorization and many more [1].

While acquisition of labeled training sample instances often requires human annotators, special devices, or expensive and slow experiments, the payload may make a fully labeled training set infeasible, whereas unlabeled sample instances is available

T. Hu (✉) · J. Yu
State Key Lab of CAD&CG, Zhejiang University, 310058 Hangzhou, China
e-mail: hutaocheng@cad.zju.edu.cn

J. Yu
e-mail: jhyu@cad.zju.edu.cn

© Springer International Publishing Switzerland 2016
R. Lee (ed.), *Software Engineering, Artificial Intelligence, Networking and Parallel/Distributed Computing 2015*, Studies in Computational Intelligence 612,
DOI 10.1007/978-3-319-23509-7_3

in large quantity and relative easy. In such situations, semi-supervised learning has proved to be great practical value. Moreover, many machine-learning researchers have found that unlabeled data, when used in conjunction with a small amount of labeled data, can produce considerable improvement in learning accuracy [2] . Thus, semi-supervised learning is also of theoretical interest in machine learning.

When encountering to the multi-class situation, semi-supervised learning comes to be more complicated [3]. A lot of recent researches of semi-supervised multi-class learning attempt to addressing those limitations. The related algorithms can be roughly divide into three categories.

1. Low-density separation. Low-density separation methods try to place boundaries in regions where there are few sample instance. One of the most commonly used algorithm is the transductive support vector machine(TSVM), TSVM attempt to labeling the unlabeled data with the decision boundary has maximal margin over all of the data, while SVM for supervised learning seeks a decision boundary with max margin between different classes. A notable advance is a multi-class extension to transductive support vector machine proposed by [4]; However, while labeled and unlabeled sample instances are treated with different measures, the related objective function is biased.
2. Boosting-based methods: There are a variety of boosting based semi-supervised multi-classification methods, these methods differ in the loss function and regularization techniques [5]. The disadvantage of them is that lake of ability to utilize the correlation between labels and input features, especially for the unlabeled data [6].
3. Graph-based methods: Graph-based methods use a graph representation for the data, with a node for each labeled and unlabeled sample instances. Some recent researches adopt Gaussian processes or Markov random walks [7], transduction by Laplacian graph [8] is also shown to be able to solve semi-supervised multi-classification problem. Although these algorithms make use of relationship between unlabeled and labeled sample instances, their computation complexity is demanding, e.g. $\mathcal{O}(n^3)$.

Aims to address the tradeoff between efficient and accuracy, and motivated by our recent work, we try to extend our multi-classifier model for semi-supervised learning. The underlying reasons are:

1 since the objective function of our model is convex, we can design efficient online algorithm with logarithmic regret, it achieve accurate solution with very little overhead. while the data is processed one after one, and the model can response to request in training stage;
2. our model is max-margin machine, which provide our model with considerable generalization capability for future unseen data.

For these reasons, we proposed an incremental model for semi-supervised learning, the model predicts a label by current state of model given unlabeled data, and then train the model with complete label, where both labeled and unlabeled data are unified under our method. The incremental model captures essences of the exploration-exploitation tradeoff.

The rest of paper is organized as follow. In Sect. 2, we introduce basics of max-margin learning. In Sect. 3, we start by describing our multi-classifier in supervised learning, convex and max-margin properties is proved in the section. In Sect. 4, we proposed a incremental algorithm for semi-supervised learning, while unlabeled data was given a prediction value, subsequent processing is the same for labeled and unlabeled data. Empirical result is present in Sect. 5. Section 6 concludes and discusses the future directions.

2 Preliminaries

We denote by $\mathcal{X} \in \mathcal{R}^d$ the sample space, $\mathcal{Y}$ the label space. For convenience, we denote sample variable by $\mathbf{x}$, and sample instances by $\mathbf{x}_t$ with subscript. Similar, label variable is denoted by $\mathbf{y}$, and label instance by $\mathbf{y}_t$.

We call the parametric family of decision function $\mathcal{H} : \mathcal{X} \times \mathcal{W} \mapsto \mathcal{Y}$ as classifiers, each classifier (given a specific parameter $\mathbf{w} \in \mathcal{W}$) takes an sample $\mathbf{x} \in \mathcal{X}$ as input and produces an output $y \in \mathcal{Y}$ which indicator which class the sample $\mathbf{x}$ belongs to. For example, in binary situation, a popular linear classifier is defined as follow.

$$h(\mathbf{x}; \mathbf{w}) = sign(\mathbf{w}^\mathsf{T}\mathbf{x}) \tag{1}$$

To get the optimal $\mathbf{w}$, we are given a training data set $\mathcal{D} = \{(\mathbf{x_1}, y_1), (\mathbf{x_2}, y_2), \cdots, (\mathbf{x_t}, y_t)\}$, we would like to find a parameter setting $\mathbf{w}$ that minimize some form of classification error. Once we have found the best parameter setting $\hat{\mathbf{w}}$, we can use the classifier to predict labels of future sample.

$$\hat{y} = h(\mathbf{x}; \hat{\mathbf{w}}) \tag{2}$$

Measure of classification error is based on loss function $\ell(\cdot)$ for each data point which depend on parameter $\mathbf{w}$ only through the classification margin, The loss function is small when the label y_t agrees with the prediction $h(\mathbf{w}; \mathbf{x})$. Also, the loss function ℓ is usually non-decreasing and convex on the margin. A regularization penalty $R(\mathbf{w})$ is also introduced in the objective function, which favors certain parameters over others (like prior).

$$\min_{\mathbf{w}, \gamma_{1:T}} \quad R(\mathbf{w}) + \frac{1}{T}\sum_{t=1}^{T} \ell(\gamma_t)$$
$$\text{s.t.} \quad y_t \cdot h(\mathbf{x_t}; \mathbf{w}) - \gamma_t \geq 0, \forall \{\mathbf{x_t}, y_t\} \in \mathcal{D} \tag{3}$$

where the symbol · denote Hadamard product(also known as Schur product or the entrywise product), unless otherwise stated, the symbol is used throughout the paper. $y_t \cdot h(\mathbf{w}; \mathbf{x})$ denote the margin, quantities γ_t works as slack variable in optimization context which represents the minimum margin that $y_t \cdot h(\mathbf{x_t}; \mathbf{w})$ must satisfy [9].

3 Generative Correlation Multi-classifier

In this section, we introduce our Bayesian multi-classifier first, and then gives out its objective function, followed by analysis of convex and max-margin property in context of supervised learning.

3.1 Graphical Model

We start by describing a generative probabilistic model correlating sample variable $\mathbf{x}$ and label variable y. The basic idea is that there is a topic space and related embedding E, the embedding E coordinates different dimension of input sample, and assign a probability measure, which aims to correlated with corresponding label variable y.

Generative correlation multi-classifier assumes following process for sample and label pairs. shown in 1

Algorithm 1 Generative Correlation Multi-Classifier

1: Choose embedding $\mathbf{w} \sim Gaussian(0, 1)$
2: **for** each i.i.d. pair $(\mathbf{x}, y)$ **do**
3:    $\mathbf{q} \leftarrow E(\mathbf{x}; \mathbf{w})$
4:    Choose $y \sim Multinomial(\mathbf{q})$
5: **end for**

We formulate the embedding E on sample variable as

$$E(\mathbf{x}; \mathbf{w}) = \phi(\mathbf{w}^\top \mathbf{x}) \tag{4}$$

parameterized by $k \times d$ matrix $\mathbf{w}$, with gaussian distribution prior. And ϕ is defined as

$$\phi(\mathbf{u} \in \mathcal{R}^k) = \frac{e^{\mathbf{u}}}{\sum_{j=1}^{k} e^{\mathbf{u}_j}} \tag{5}$$

3.2 Objective Function

Suppose we have a set of pair instances $\{(\mathbf{x_t}, y_t)\}$ generated by our model, the joint distribution is then given by:

$$p(\{\mathbf{x}_t, y_t\}_{t=1}^T, \mathbf{w}) = p(E) \prod_{t=1}^T p(\mathbf{q}_t | \mathbf{w}, \mathbf{x_t}) p(y_t | \mathbf{q}_t) \tag{6}$$

Taking *log* operator on joint distribution, we get an log-likelihood form objective function.

$$\max_{\mathbf{w}} \mathcal{L}(\mathbf{w}; \{\mathbf{x}_t, y_t\}_{t=1}^T) = -\frac{1}{2} \|w\|_{\mathcal{F}}^2 + \sum_{t=1}^T \langle y_t, \log \phi(\mathbf{w}^\top \mathbf{x}_t) \rangle \tag{7}$$

where $\| \cdot \|_{\mathcal{F}}$ denote Frobenius norm of matrix.

3.3 Convex Analysis

The logic of convex analysis is based on requirement of online convex optimization, for an objective function consisting of two part, regularization term and data term, an online algorithm with logarithmic regret is proposed if we can prove the convexity of data term [10, 11].

For analysis purpose, we list a minimization optimization problem which is equivalent to 7.

$$\min_{\mathbf{w}} \frac{1}{2} \|w\|_{\mathcal{F}}^2 - \sum_{t=1}^T \langle y_t, \log \phi(\mathbf{w}^\top \mathbf{x}_t) \rangle \tag{8}$$

where $\|w\|_{\mathcal{F}}^2$ act as regularization term, then we need to prove the following proposition.

Proposition 1 $\langle y, \log \phi(\mathbf{w}^\top \mathbf{x}) \rangle$ *is concave on* $\mathbf{w}$

Proof We start by noting property of

$$\log \phi(\mathbf{u}) = \mathbf{u} - \log \sum_k e^{\mathbf{u}_k} \tag{9}$$

there is a common factor $\log \sum_k e^{\mathbf{u}_k}$, taking partial derivative on $\log \sum_k e^{\mathbf{u}_k}$ related to $\mathbf{u}$, the first order derivative form is formulated as

$$d \log \sum_k e^{\mathbf{u}_k} = \langle \phi, d\mathbf{u} \rangle \tag{10}$$

The second order derivative is by analogy calculas

$$d^2 \log \sum_k e^{\mathbf{u}_k} = \phi_i(1 - \phi_j) du_i du_j \tag{11}$$

the Hessian matrix is sum of two positive rank one matrix $\phi(1 - \phi)^\top + (1 - \phi)\phi^\top$, then $\log \sum_k e^{\mathbf{u}_k}$ is convexity. Cause the other part is linear and y is not less than zero, we get the conclusion that $\langle y, \log \phi(\mathbf{w}^\top \mathbf{x}) \rangle$ is concave.

Moreover, by noting the embedding assign a probability measure for sample instance, embedding act as prior of probabilities, which corresponds to dirichlet parameter [12] in exponential family. And there is translation invariance property of ϕ, $\phi(\mathbf{u}) = \phi(\mathbf{u} + s)$, $\mathbf{u} \in \mathcal{R}^k$, $s \in \mathcal{R}$. Besides, if we define functions $\{g_t\}_{t=1}^T$ on $\mathbf{w}$ as:

$$g_t(\mathbf{w}) = -\langle y_t, \log \phi(\mathbf{w}^\top \mathbf{x}_t) \rangle \tag{12}$$

then we can get the dual form of optimization problem 8.

$$\max_{\mu_1, \cdots, \mu_T} \quad -\left\{ \frac{1}{2} \| \sum_t \mu_t \|_\mathcal{F}^2 + \sum_t g_t^*(\mu_t)) \right\} \tag{13}$$

given $\mathbf{w}^*$ is the solution of 8, and $\{\mu_t^*\}_{t=1}^T$ is the solution of its dual problem 12, then we know there is relationship

$$\mathbf{w}^* = -\sum_{t=1}^T \mu_t^* \tag{14}$$

For these reasons, we design an online learning algorithm aggregating dualities without averaging, which we terms it Duality Aggregation, shown in 2.

Algorithm 2 Duality Aggregation for Generative Correlation Multi-Classifier

---

Input: Training data $\mathcal{D} = \{(\mathbf{x}_t, y_t)\}_{t=1}^T$
Output: Embedding E^* with parameter $\mathbf{w}^*$
1: $\mathbf{w}_0 \leftarrow 0$
2: **for** $t = 1$ to T **do**
3:     $\mathbf{w}_t \leftarrow \mathbf{w}_{t-1} + (y_t - \phi(\mathbf{w}_{t-1}^\top \mathbf{x}_t))\mathbf{x_t}^\top$
4: **end for**
5: $\mathbf{w}^* \leftarrow \mathbf{w}_T$

3.4 Max-Margin Property Analysis

In the previous section, we know margin and loss function is key elements in max-margin learning related optimization problem, we try to build the concepts in our optimization problem. By noting that inner product $\langle y_t, \log \phi(\mathbf{w}^\top \mathbf{x}_t) \rangle$ is not greater than zero, we denote its Hadamard product form as margin, and also we can define loss function $\ell(\gamma) = -\gamma$, it is obvious convex and non-decreasing. With the notation margin and loss function introduced, 7 can be expressed as

$$\max_{\mathbf{w}} -(\frac{1}{2}\|w\|_{\mathcal{F}}^2 + \sum_{t=1}^{T} \ell(\overbrace{y_t \cdot \log \phi(\mathbf{w}^\top \mathbf{x}_t)}^{\text{margin}})) \tag{15}$$

Then, we show the follow proposition.

Proposition 2 *Our model is max-margin machine*

Proof We know that, a machine is said to be max-margin if and only if its learning objective function has following form:

$$\min_{\mathbf{w}} \quad \frac{1}{2}\|w\|^2 + \lambda \sum_t \ell(\gamma_t)$$
$$\text{s.t.} \quad y_t \cdot h(\mathbf{w}^\top \mathbf{x_t}) \geq \gamma_t, \forall t \in [T] \tag{16}$$

It is sufficient to show that 16 hold by introducing lower bound variables $\{\gamma_t\}_{t=1}^{T}$ and replacing the margins $\{y_t \cdot \log \phi(\mathbf{w}^\top \mathbf{x}_t)\}_{t=1}^{T}$ in 15.

Before further step, we recall one more piece of definition from convex analysis. The Fenchel conjugate of a function $f : S \mapsto \mathcal{R}$ is defined as

$$f^*(x^*) = \sup_{x \in S} \langle x^*, x \rangle - f(x)$$

which corresponds to a optimization problem.

Fenchel conjugate has a nice property, if f is closed and convex, then the Fenchel conjugate of f^* is f itself (a function is closed if for all $\alpha > 0$ the level set $\{x : f(x) \leq \alpha\}$ is a closed set).[13, 14]

Let's back to proof of necessary condition. Consider the optimization problem 16, the inequality constraints should be equality ones, otherwise the objective is not optimal for nondecreasing properties of ℓ

$$\min_{\mathbf{w}} \quad R(\mathbf{w}) + \sum_t \ell(\gamma_t)$$
$$\text{s.t.} \quad y_t \cdot h(\mathbf{x_t}; \mathbf{w}) - \gamma_t, \forall \{\mathbf{x_t}, y_t\} \in \mathcal{D} \tag{17}$$

introducing T vectors $\lambda_1, \cdots, \lambda_T$, each λ_t is Lagrangian multiplier of the equality $\langle y_t, h(\mathbf{x_t}; \mathbf{w}) \rangle = \gamma_t$.then we obtain the following Lagrangian.

$$L(\mathbf{w}, \gamma_1, \cdots, \gamma_T, \lambda_1, \cdots, \lambda_T)$$
$$= R(\mathbf{w}) + \sum_t [\ell(\gamma_t) + \langle \lambda_t, y_t \cdot h(\mathbf{x_t}; \mathbf{w}) - \gamma_t \rangle] \tag{18}$$

applying Fenchel duality theorem [13] on γ and λ in turn.

$$\min_{\mathbf{w}, \gamma_1, \cdots, \gamma_T} \max_{\lambda_1, \cdots, \lambda_T} L(\mathbf{w}, \gamma_1, \cdots, \gamma_T, \lambda_1, \cdots, \lambda_T)$$
$$= \min_{\mathbf{w}} \max_{\lambda_1, \cdots, \lambda_T} R(\mathbf{w}) + \sum_t \left[-\ell^*(\lambda_t) + \langle \lambda_t, y_t \cdot h(\mathbf{x_t}; \mathbf{w}) \rangle \right]$$
$$= \min_{\mathbf{w}} R(\mathbf{w}) + \sum_t \ell(y_t \cdot h(\mathbf{x_t}; \mathbf{w}))) \tag{19}$$

where ℓ^* is Fenchel conjugate of ℓ, we did maximization and minimization on $\{\gamma_1, \cdots, \gamma_T\}$ and $\{\lambda_1, \cdots, \lambda_T\}$ sequentially, then we obtain a unconstrained optimization problem again.

The proposition leads to that our model has similar generalization bound to SVM based multi-classifiers. Moreover, cause local variables $\{\gamma_t\}_{t=1}^T$ are eliminated, there is only one optimization variable $\mathbf{w}$ in objective function 7, which is much simple than the optimization problem with constraints 16.

4 Extending Our Model to Semi-Supervised Learning

In the previous section, we proved that our model has two notable properties. Convex property indicates that we can design an efficient online algorithm with logarithmic regret, while the data is processed one after one, we can pause the learning procedure, and continue the learning procedure after responding to prediction request without any effect on learning. Max-margin property indicates our model has comparable generalization capability as other max-margin machines, e.g. Support Vector Machines (SVMs), prediction on future unseen data can achieve nearly the same performance as in training stage.

For these reasons, we propose an incremental algorithm for semi-supervised learning, when given unlabeled data, the algorithm pause the training procedure, predicts a label with current state of model, and then continue to train the model with complete label data.

Label prediction is defined as following optimization problem based on max-entropy principle

$$\max_{y_t \in \mathcal{Y}} \langle y_t, \log \mathbf{q}_t \rangle - \langle y_t, \log y_t \rangle \qquad (20)$$

where $\mathbf{q}_t$ is the probability measure embedding E assigning to input sample $\mathbf{x}_t$,

$$\mathbf{q}_t = \phi(\mathbf{w}^\top \mathbf{x}_t) \qquad (21)$$

the objective function consists of two part, cross entropy $\langle y_t, \log \mathbf{q}_t \rangle$ and entropy of $y_t : \langle y_t, \log y_t \rangle$, representing confidence of y_t given $\mathbf{q}_t$. The solution is

$$y_t^* = \mathbf{1}(\cdot = \arg\max(\mathbf{q}_t)) \qquad (22)$$

where $\mathbf{1}(\cdot = \arg\max(\mathbf{q}_t))$ is indicator function, which means dimension with max value of $\mathbf{q}_t$ labeled with 1, others labeled with 0.

With the proposed algorithm, both labeled and unlabeled data are unified under our method. The incremental algorithm captures essences of the exploration-exploitation tradeoff. The algorithm shown in 3.

Algorithm 3 Incremental Duality Aggregation for Semi-Supervised Multi-Class Learning

---

Input: Training data $\mathcal{D} = \{(\mathbf{x}_t, y_t)\}_{t=1}^T$
Output: Embedding E^* with parameter $\mathbf{w}^*$
1: $\mathbf{w}_0 \leftarrow 0$
2: **repeat**
3:   **for** $t = 1$ to T **do**
4:     **if** $\mathbf{x}_t$ is unlabeled **then**
5:       $y_t \leftarrow \mathbf{1}(\cdot = \arg\max(\phi(\mathbf{w}^\top \mathbf{x}_t)))$
6:     **end if**
7:     Computation aggregate with
8:     $\mathbf{w}_t \leftarrow \mathbf{w}_{t-1} + (y_t - \phi(\mathbf{w}_{t-1}^\top \mathbf{x}_t))\mathbf{x_t}^\top$
9:   **end for**
10: **until** Convergence
11: $\mathbf{w}^* \leftarrow \mathbf{w}_T$

---

5 Experiments

We present empirical results to demonstrate prediction accuracy and converge rate of proposed semi-supervised learning algorithm, the result demonstrate merits inherited from both online convex optimization and max-margin learning. Data set is divided into training and test samples, we feed the model with training samples in

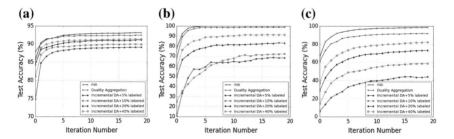

Fig. 1 Classification accuracy versus iteration numbers. **a** MNIST. **b** COIL20. **c** COIL100

format $(\mathbf{x}, y)$ (labeled sample) or only $\mathbf{x}$ (unlabeled sample), after training, we feed the model with $\mathbf{x}_t$ in test samples set, the model returns $\hat{y}_t$, prediction accuracy is defined as $\frac{\sum_t 1(\hat{y}_t = y_t)}{|\text{Test Samples}|}$ representing the probability whether prediction label $\hat{y}_t$ is equals to true label y_t. The performance mainly depend on proportion of labeled samples in training data, we test four proportions: 5, 10, 20 and 40 %, governed by exponential relationship. We care about dynamic evolutional performance of algorithms, prediction accuracy of various iterations are evaluated.

To order to estimate limits of semi-supervised learning, supervised learning using Duality Aggregation algorithm is introduced, also training with whole samples (not only training data, test data also included) is introduced, which represents the true risk of supervised learning.

We evaluate performance on three data set MNIST COIL20 and COIL100, which have been extensively evaluate in context of multi-class learning.

5.1 MNIST

The MNIST database[1] is handwritten digits ranging from 0 to 9, consist of 60000 training examples and 10000 test examples, The size of each image is 28×28 pixels, with 256 gray levels per pixel, thus each image is represented by a 784-dimensional vector.

Performance is reported in Fig. 1a and Table 1, we can see that, with only 5 % labeled data, our algorithm can achieves 89.10 ± 0.20 prediction accuracy after 20 iterations, there are less than 5 % accuracy lost comparing with supervised learning (92.47 ± 0.05 %) and true risk (93.18 ± 0.05 %).

---

[1] http://yann.lecun.com/exdb/mnist/.

Table 1 Accuracy of incremental duality aggregation after 20 iterations

Dataset\Methods	Incremental duality aggregation with labeled data(%)				Duality aggregation (%)	Risk (%)
	5 (%)	10 (%)	20 (%)	40 (%)		
MNIST	89.10 ± 0.20	89.90 ± 0.13	91.12 ± 0.15	91.35 ± 0.12	92.47 ± 0.05	93.18 ± 0.05
COIL20	68.24 ± 0.50	72.29 ± 1.17	82.97 ± 1.51	91.40 ± 0.82	98.76 ± 0.35	100 ± 0
COIL100	43.91 ± 1.47	58.65 ± 0.89	73.07 ± 0.60	82.31 ± 0.54	91.85 ± 0.33	98.10 ± 0.25

5.2 COIL20 and COIL100

COIL20[2] contains 20 objects, the images of each objects were taken 5 degrees apart as the object is rotated on a turntable and each object has 72 images, The size of each image is 32×32 pixels, with 256 gray levels per pixel, thus each image is represented by a 1024-dimensional vector. COIL100[3] contains 100 objects, other setting is similar to COIL20.

Cause the data set is not divided into training and test samples, we divide the data set into training samples and test samples randomly with ratio 7 : 3, performance is shown in Fig. 1b, c and Table 1. We can see that with 5 % labeled data, prediction accuracy achieves $68.24 \pm 0.50\%$ and $43.91 \pm 1.47\%$ in COIL20 and COIL100 separately. The performance is getting better with the increasing of labeled data proportion, while it can achieve $91.40 \pm 0.82\%$ and $82.31 \pm 0.54\%$ with 40 % labeled data. Comparing with the performance of MNIST, the reason of the unperfect performance may caused by inadequate learning, as there are 72 samples for each class.

6 Conclusions

In this paper, we proposed an incremental model for semi-supervised learning, the model use two notable properties:

1. convex of objective function. we can design efficient online algorithm with logarithmic regret, it achieve accurate solution with very little overhead. moreover, while the data is processed one after one, it can also response to prediction request in training stage;
2. max-margin learning, which provide our model with considerable generalization capability for future unseen data.

As unlabeled sample is given a prediction value, then train the model with complete label. both labeled and unlabeled data are unified in our method, our incremental model capture essences of the exploration-exploitation tradeoff. The empirical study shows our approach has acceptable results. Our future work will focus on applying the approach to more broader scenarios.

Acknowledgments This work is supported by the National Natural Science Foundation of China (No.61379069) and the Key Technologies R&D Program of China (No.2014BAK09B04).

---

[2]http://www.cs.columbia.edu/CAVE/software/softlib/coil-20.php.
[3]http://www.cs.columbia.edu/CAVE/software/softlib/coil-100.php.

References

1. Daniely, A., Shalev-Shwartz, S.: Optimal learners for multiclass problems, *arXiv preprint* arXiv:1405.2420 (2014)
2. Zhu, X., Goldberg, A.B.: Introduction to semi-supervised learning. Synth. Lect. Artif. Intell. Mach. Learn. **3**(1), 1–130 (2009)
3. Valizadegan, H., Jin, R., Jain, A.K.: Semi-supervised boosting for multi-class classification. In: Machine Learning and Knowledge Discovery in Databases, pp. 522–537, Springer, Berlin(2008)
4. Xing, E.P., Jordan, M.I. Russell, S., Ng, A.Y.: Distance metric learning with application to clustering with side-information. In: Advances in neural information processing systems, pp. 505–512, (2002)
5. Saffari, A., Leistner, C., Bischof, H.: Regularized multi-class semi-supervised boosting. In: IEEE Conference on Computer Vision and Pattern Recognition, 2009. CVPR 2009, pp. 967–974 (2009)
6. Wang, B., Tu, Z., Tsotsos, J.K.: Dynamic label propagation for semi-supervised multi-class multi-label classification. In: 2013 IEEE International Conference on Computer Vision (ICCV), pp. 425–432 (2013)
7. Azran, A.: The rendezvous algorithm: Multiclass semi-supervised learning with markov random walks. In: Proceedings of the 24th international conference on Machine learning, pp. 49–56, ACM, (2007)
8. Goldberg, A., Recht, B., Xu, J., Nowak, R., Zhu, X.: Transduction with matrix completion: Three birds with one stone. In: Advances in neural information processing systems, pp. 757–765, (2010)
9. Jebara, T.: *Machine Learning: Discriminative and Generative*. Kluwer Academic, Boston (2004)
10. Shalev-Shwartz, S., Kakade, S.M.: Mind the duality gap: Logarithmic regret algorithms for online optimization. In: Advances in Neural Information Processing Systems, pp. 1457–1464 (2009)
11. Srebro, N., Sridharan, K., Tewari, A.: On the universality of online mirror descent. In: Advances in Neural Information Processing Systems, pp. 2645–2653 (2011)
12. Blei, D.M., Ng, A.Y., Jordan, M.I.: Latent dirichlet allocation. J. Mach. Learn. Res. **3**, 993–1022 (2003)
13. Boyd, S., Vandenberghe, L.: *Convex Optimization*. Cambridge university press, Cambridge (2004)
14. Shalev-Shwartz, S.: Online Learning: Theory, Algorithms, and Applications, Ph.D. thesis, Hebrew University of Jerusalem (2007)

Improving Hypervisor Based SSD Caching with Logically Partitioned Blocks and Scanning in Cloud Environment

Hee Jung Park, Kyung Tae Kim, Byungjun Lee, Rhee Man Kil and Hee Yong Youn

Abstract In the era of big data and cloud computing the virtual machine (VM) environment is important where multiple VMs of different operating system and application can be simultaneously run on the same host. In the VM environment the conventional hard disk drive (HDD) has limitations such as low random access performance and high power consumption. Solid State Drive (SSD) is an emerging storage technology, playing a critical role in revolutionizing the storage system design. Recently, SSD storage caching is widely studied for VM-based systems. The existing works on cache space allocation identify the space demand of each VM based on hit ratio. They are not effective for the VMs of shared SSD cache due to the filtering effect of higher-level caches. In this paper we propose a novel hypervisor-based SSD caching scheme, employing a new metric to accurately determine the demand on SSD cache space of each VM. Computer simulation confirms that it substantially improves the accuracy of cache space allocation compared to the existing schemes. It also allows to display comparable hit ratio as the existing schemes with less amount of SSD cache for the VMs.

Keywords Cache allocation · SSD caching · Virtual machine · Hit ratio · Hypervisor

H.J. Park · K.T. Kim · B.J. Lee · R.M. Kil · H.Y. Youn (✉)
College of Information and Communication Engineering, Sungkyunkwan University,
2066, Seoburo, Suwon 440-746, Republic of Korea
e-mail: youn7147@skku.edu

H.J. Park
e-mail: hjhjpark@skku.edu

K.T. Kim
e-mail: kyungtaekim76@skku.edu

B.J. Lee
e-mail: byungjun@skku.edu

R.M. Kil
e-mail: rmkil@skku.edu

© Springer International Publishing Switzerland 2016   45
R. Lee (ed.), *Software Engineering, Artificial Intelligence, Networking
and Parallel/Distributed Computing 2015*, Studies in Computational Intelligence 612,
DOI 10.1007/978-3-319-23509-7_4

1 Introduction

In recent years virtualization technology enables multiple virtual machines (VMs) to be run on a physical machine, where each VM is run independently on its own operating system. Virtualization technology has been adopted in various IT fields because of its ability to improve the utilization of hardware resource, achieving low-power consumption, simplifying server management, and reducing maintenance cost. In a typical VM environment, multiple VMs are simultaneously run on the same host.

In VM environment, high-performance storage systems are in high demand especially for data-intensive computing. However, most storage systems, even those specifically designed for high-speed data processing, are still built on conventional hard disk drive (HDD) with several long-lasting limitations including low random access performance and high power consumption. Unfortunately, these problems essentially stem from the mechanical nature of HDD, and thus are difficult to be addressed via mechanical solution. Flash memory-based Solid State Drive (SSD) is an emerging storage technology which plays a critical role in revolutionizing the design of storage system. Different from HDD, SSD are completely built on semiconductor chips without any moving parts. Such fundamental difference makes SSD capable of providing one order of magnitude higher performance than HDD, and lets it be an ideal storage medium for building high-performance shared storage systems. Due to these advantages SSD caching has been widely studied in conventional systems [1, 2]. However, it would be impractical to directly apply the existing solutions to the VM systems because the SSD caching scheme designed for a VM system must maximize the utilization of shared SSD cache while ensuring performance isolation among the VMs.

The previous works on cache partitioning for VMs focus on the identification of the space demand on cache of each application [3]. For example, some researches proposed to monitor the number of hits/misses of each cache unit and then use the data as a basis for computing the space demand [4]. Some studies proposed to use the change of hit ratio at main memory level during a time window as a metric to predict the space demand [5]. However, the hit ratio-related techniques cannot be effective for shared cache due to the filtering effect of higher-level caches. For this reason, a cache space reallocation scheme based on random sampling was proposed in [6]. The space allocation for a VM is determined by the memory utilization and maximum/minimum memory quota predefined by the system administrator. The VM of lower memory utilization has a smaller share reclaimed and thus it is more likely to get memory. In this case, however, it can predict the memory requirement of a VM only when there is available memory. In addition, the prediction is inaccurate because of the mechanism of random scanning of the blocks.

In this paper a hypervisor-based SSD caching scheme is proposed, which effectively manages the cache in the multi-VM environment by collecting and exploiting the runtime information from both the VMs and storage devices. Due to the unique position of hypervisor between the VMs and hardware devices, it does not require any modification of guest OS, user applications, or the underlying storage systems.

The proposed scheme uses a metric called "*HLP*" to identify the cache space demand of each VM at runtime. In essence, *HLP* is the ratio of the cache being effectively used by a VM to the total cache space allocated to it. It is a critical reference for cache space allocation. Computer simulation reveals that the proposed scheme improves accuracy and more effectively uses the cache space compared with the existing schemes.

The rest of the paper is organized as follows. In Sect. 2 the existing schemes related to hypervisor-based SSD caching are explained. The proposed scheme is presented in Sect. 3, and Sect. 4 compares its performance with the existing methods using computer simulation. Finally, Sect. 5 presents the conclusion and future work.

2 Related Works

2.1 Hypervisor-Based SSD Caching

Recently, various hypervisors or VM monitors are run on top of physical machines which schedule the execution of VMs. Here multiple instances of operating systems may share the virtualized hardware resources. Among them, Xen hypervisor is a popular abstraction layer existing between the guest domains and physical hardware, and it is responsible for resource allocation and isolation. Here the LRU (Least Recently Used) policy is usually employed to remove data from the cache. Some studies track the count of block accesses to identify frequently used blocks, and then cache them in SSD [5, 7].

The caches come in two varieties. Firstly, dedicated deployment such as memcached where each node is allocated a fixed amount of memory along with opportunistic caches such as Linux buffer and page cache that can expand to consume underutilized memory. With opportunistic cache a process of a VM may greedily consume the memory pages if there is no other process inside the VM using the memory. However, there might be another VM of higher priority on the same host which could make better use of the memory. Then efficient use of virtual resources becomes difficult, and a new approach needs to be adopted. Meanwhile, dedicating a fixed memory region to a memcached node is convenient for offering a predictable level of quality of service. Also, UniCache allows flexible allocation of storage resource to the operating system and applications by offering a unified caching service at the hypervisor level as shown in Fig. 1.

Here data are split between hypervisor controlled main memory and flash memory to provide varying levels of performance based on the application type and priority of the VM. Expanding the cache to include both memory and SSD allows a much larger amount of data to be stored, which is very important in virtualized environment where competing VMs need to make efficient use of limited memory resources [8–10].

There exist three different approaches for using SSD cache in VM environment. Firstly, SSD is directly managed by the hypervisor, where the management center is located between the VMs and hardware resources. The VM-based SSD caching

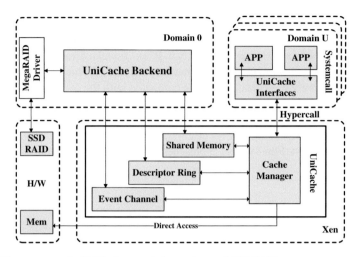

Fig. 1 The structure of a UniCache store in hypervisor and SSD RAID

has significant disadvantages such that the guest OS or user application needs to be modified to manage the cache. This incurs extra burdens on the users, and is hardly available for legacy systems. The storage-based SSD caching shows a drawback in its isolated design approach. The block interface between the storage system and the virtualization software stack is primitive without the ability to deliver rich semantic information. Local optimization in the storage subsystem may not enhance the performance of overall VM system.

This paper employs hypervisor-based SSD caching to avoid the limitations in VM-based and storage-based SSD caching, while retaining their advantages. The hypervisor can manage SSD cache for the VMs in a transparent way, addressing the problem of modifying guest OS or application. Here the VM activities, particularly I/O requests are managed by the hypervisor which collects critical information required for effective management of SSD cache. With the full access privilege to hardware resources, the hypervisor can directly enforce the space allocation decisions to maximize resource utilization in an efficient way [6, 11, 12].

2.2 Cache Partitioning

At present, a number of researchers have proposed flash-based buffer management algorithm for SSD such as CFLRU [13] and improved CRLRU [14]. Taking advantage of the asymmetry in flash read-write performance, CFLRU is a kind of buffer replacement strategy which first replaces read-only pages and assumes that write cost of flash is far greater than read cost. Its key idea is dividing LRU linked list into two parts: working area and replacement area. Once cache is full and some data need to be replaced to outside, CFLRU chooses read-only pages for replacement according to LRU supposing that there exists read-only data in the replacement area. When there are only dirty pages in the replacement area, the ones at the tail of the linked list are replaced. Other researchers improved the traditional LRU and LFU policy to accommodate diverse requirements of the applications [15–19].

Note that there exists a high potential for SSD to be widely employed in large-scale cluster storage system. SSD is more expensive than traditional HDD, but performs better for random reads and writes. S-CAVE [6] is a flash cache partitioning scheme which tries to maximize cache utilization for multiple VMs on a single VMware host by running a hypervisor module. Based on the identification of runtime working set, S-CAVE monitors the changes in locality, especially transient bursts in data reuse. Here the performance of caching is measured by the number of cache hits an application encounters. If proper data blocks are cached, the number of cache hits will increase, and accordingly the effectiveness of caching. Therefore, an efficient algorithm needs to be employed to maximize the number of cache hits.

Figure 2 illustrates the structure of S-CAVE. For each VM, S-CAVE launches a module, called Cache Monitor, to manage the allocated cache space and keep the cache transparent to the VM. In order to effectively allocate the shared SSD cache space among multiple VMs, S-CAVE also uses a central control, called Cache Space Allocator, to analyze the information on cache usage collected from each cache monitor and make the decision on cache allocation accordingly.

The key idea of S-CAVE is to effectively allocate an appropriate amount of cache space to each VM. For this, S-CAVE identifies the demand of cache space of each VM, and each cache monitor is required to provide accurate information on the demand of the SSD cache space of the VM it monitors. Then cache space allocation is made considering the demands of all VMs. While satisfying the demand on the cache of

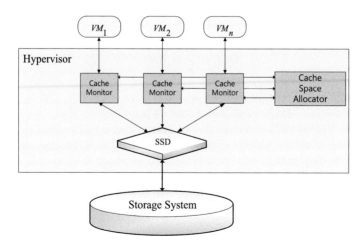

Fig. 2 The structure of the S-CAVE scheme

each VM represented by the ratio of used cache space, the cache space allocation of all VMs needs to be properly balanced. The proposed scheme efficiently resolves this issue as shown next.

3 The Proposed Scheme

The environment of VM is highly dynamic, where multiple VMs of different and changing cache demands are run on a host. To rapidly reflect the runtime dynamics and guarantee effective and fair sharing of cache space, a dynamic control mechanism is proposed to periodically cross-compare the cache demand of each VM and adjust the space allocation accordingly. Through this, the VMs of increasing demand is granted more cache space, while some portion of already allocated cache space of those of decreasing demand is deprived. In order to achieve fine adjustments, the previous decisions are also taken into account as a feedback when a new decision is made.

The proposed scheme consists of two steps. The first step is to estimate the value of *HLP* identifying the demand of each VM on the cache space. Cache space reallocation is made in the next step, considering the demands of all the VMs.

3.1 Assessment of Cache Utilization

This paper proposes a new metric called *HLP* which is the ratio of the size of cache space being used to the allocated cache space during a time window. For a specific VM for which n blocks have been allocated, if m unique cache blocks have been accessed within a time window, then *HLP* is obtained by Eq. (1).

$$HLP = \frac{m}{n} \times 100\,\% \qquad\qquad (1)$$

To accurately and efficiently estimate the *HLP* value at runtime, two counters, C_i and C_i^d, are manipulated for each VM, where C_i is the total number of cache blocks allocated to VM_i and C_i^d is the number of unique cache blocks used by VM_i. At the beginning of a time window, both counters are set to 0, and the metadata of the blocks residing in the global pool is scanned to update the counters. Since the global pool contains all the blocks allocable to different VMs, only the counter corresponding to the accessed block for the VM is incremented by one. Whenever, C_i^d is incremented, C_i is also incremented by one. Figure 3 shows how the proposed counters are manipulated.

In order to obtain accurate *HLP* value, all reference counters need to be scanned for the given time window. A small time window enables quick adjustment of cache allocation, but it is impossible to complete the scan of entire counters during the

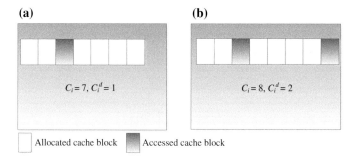

(a)   **(b)**

$C_i = 7, C_i^d = 1$ $C_i = 8, C_i^d = 2$

☐ Allocated cache block   ▨ Accessed cache block

Fig. 3 An example of counter manipulation for VM_i

window. A large time window allows to finish the scan, but cache allocation becomes less responsive to the runtime dynamics. As a result, selecting an appropriate window size is an important issue in achieving the best performance.

In order to efficiently estimate the *HLP* value while rendering a high accuracy, a new sampling mechanism is adopted in this paper. Here a time window is split into multiple small sampling periods, and an idle period is inserted between two consecutive sampling periods. This approach is to reduce the computation overhead while increasing the effectiveness of the scan. Note that the change in the access during the idle period can be counted in the subsequent sampling period, which increases the accuracy of *HLP*. Within each sampling period, the scanning begins from the block where the previous scan ends. The scanning operation stops when the sampling period expires. Figure 4 shows an example of the operation of the proposed scheme obtaining the *HLP* values. Note here that the physically tied SSD is logically partitioned, and the demand on cache space of each partition is monitored during each time window. The proposed scheme is different from the existing ones in the management of the scan which allows accurate estimation of the demand on the cache for each VM and cache space allocation based on it.

The value of *HLP* for *j*th partition of *i*th time window, HLP_i^j, having k sampling periods is the average value of k samples, $Sample_i^j$ ($i = 1,…,k$) which can be formulated by Eq. (2).

$$HLP_i = \frac{1}{k} \sum_{i=1}^{k} Sampling_i \qquad (2)$$

Assume that current time window is i. HLP_i denotes the final *HLP* of *i*th time window covering all the partitions. Then,

$$HLP_i = \frac{1}{p} \sum_{j}^{p} HLP_i^j \qquad (3)$$

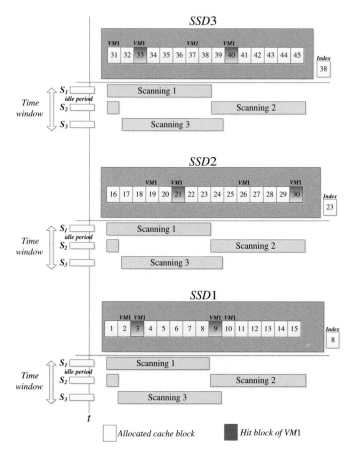

Fig. 4 An example of scanning for the estimation of *HLP*

where p is the number of partitions. In the example of Fig. 4 the SSD is partitioned into three parts, SSD1, SSD2, SSD3, and the time window consists of three scanning periods. Each partition contains 15 blocks, and VM_1 is allocated 12 blocks. The '*index*' in each SSD partition points the block number scanned last in the previous sampling period. In Fig. 4, it is assumed that S_1 has just been finished.

To further improve the accuracy, the current *HLP* value is averaged with the two recent *HLP* values using a weight parameter a, enabling small time window to be more responsive to the change in the space demand. The final HLP_i is obtained by Eq. (4). In this paper a is assumed to be 0.8. Observe from the figure that six blocks were accessed out of 12 blocks for VM_1, and thus its *HLP* in this time window is 0.5.

$$HLP_i = HLP_i^j \cdot a + (HLP_{i-2} + HLP_{i-1})/2 \cdot (1 - a) \tag{4}$$

Algorithm 1 shows the procedure for the estimation of *HLP*.

Algorithm 1 Estimation of *HLP*

1: //Initially setting
2: Index block = 0
3: Sequential start block = Index block
4: Totally running *VM*s = *N*
5: Allocated blocks of VM_i = VMC_i
6: Hit blocks of VM_i = VMC_i^d
7: Number of partitions = *p*
8:
9: **for** each VM_i ($1 \leq i < N$) **do**
10:   $VMC_i = 0$
11:   $VMC_i^d = 0$
12: **end for**
13:
14: //Calculate *HLP* of a time window
15: **for** each *p* **do**
16:   **for** $S_1 \leq S_k < k$ **do**
17:      **if** Scanned block = Allocated block
18:         VMC_i ++
19:      **else if** Scanned block = Hit block
20:         VMC_i^d ++
21:      **end if**
22:      $HLP = VMC_i^d / VMC_i + VMC_i^d * 100$
23:      $HLP_i^j += HLP / k$
24:      Index block = Last seek block
25:   **end for**
26:   $HLP_i = HLP / p$
27: **end for**
28:
29: $HLP_i = HLP_i^j * a + (HLP_{j-2} + HLP_{j-1}) / 2 * (1-a)$

3.2 Allocation of Cache

In cloud computing the available cache space is allocated to the cloud applications. The cache space is provided on demand in a fine-grained, multiplexed manner. Here the cache space allocation is based on the infrastructure as a service (IaaS).

Resource fragmentations occur as the resources are continuously allocated and deallocated. Even though the entire amount of available resource is enough to satisfy the need, it cannot be allocated to the requesting application due to fragmentation. The proposed cache allocation scheme considers the issue of scarcity of resources because the resources are usually limited while the demand is high in the hypervisor-based cloud environment. A dynamic allocation scheme is thus adopted to solve the problem [20].

Each time the cache space of a VM is adjusted, the amount of change is determined by the parameter *AlloCache*, which is the average number of blocks the VM can access within a time window. It is obtained by averaging the number of accesses in the previous time windows. VM_{min} and VM_{max} denote the VM of the smallest *HLP* and largest *HLP*, respectively. In other words, *AlloCache* of VM_{min} is the rate of missed accesses for a VM in the recent time window. The purpose of using the parameter is to ensure that the new data accessed by VM_{max} in the subsequent time window can be accommodated using the new available cache space released by VM_{min}. The proposed scheme finds VM_{max} to increase its cache space, and the amount of increase is determined by *AlloCache* of VM_{min}. Algorithm 2 below explains how to allocate the cache space. In case the total free space is smaller than 5%, the cache space of VM_{min} is swapped with that of VM_{max} for maximizing the utilization of the total space of SSD [21]. Additionally, the proposed scheme applies the CLOCK-based cache replacement approach to be more adaptive [22]. The CLOCK algorithm captures the information on cache access and exploits the frequency of cache access via the reference bit unlike LRU [23].

In Algorithm 2, the dynamic cache allocation scheme based on *HLP* is presented. With the *HLP* identifying the demand on the cache of a VM, the proposed scheme effectively balances the allocation between the VMs considering the availability of hardware resources. For this, the configuration information including the amount of allocated cache of the VMs is utilized. Here Cache(VM_i) denotes the amount of cache allocated to VM_i.

Algorithm 2 Allocation of Cache Space

1: Totally running $VMs = N$
  2: **if** $N = 1$ **then**
    3:    Cache $(VM_0) = \infty$
    4:    exit
  5: **end if**
  6:
  7: **if** Total free space > 5% **then**
    8:   **for** VM_i $(1 \leq i \leq N)$ **do** VM Cache (VM_i) = upper limit
  9:   **end for**
  10: **end if**
  11:
  12: Calculate *HLP*
  13: **Sort** { $HLP(VM_i) \mid 1 < i < N$} }
  14:
  15: **if** Total free space < 5% **then**
    16:   **Swap**(Cache (VM_{max}), Cache (VM_{min}))
  17: **end if**
  18:
  19: Cache $(VM_{max}) + = AlloCache (VM_{max})$
  20: Cache $(VM_{min}) - = AlloCache (VM_{min})$

---

4 Performance Evaluation

In this section the performance of the proposed scheme is evaluated which uses *HLP* to correctly identify the demand on the cache of the VMs and thereby reallocate the cache space when managing multiple VMs. The simulation was conducted with a PC of Intel 3.5Ghz i3 CPU, 16GB RAM, 64bit Window 7, and the simulation code was written in C++ language with Visual studio 2010. The size of SSD cache was gradually increased from 1,000 to 10,000 blocks for the evaluation. The effectiveness of the proposed scheme is compared with that of S-CAVE in terms of cache space utilization and hit ratio. In the simulation the length of time window, sampling period, and idle period are set to be 1 sec, 0.2 sec, 0.1 sec, respectively.

4.1 Accuracy

First, one VM is used to run a single workload each time. The VM starts with a small size cache which will then be dynamically adjusted during runtime. Figure 5 compares the accuracy of the proposed scheme with S-CAVE. The accuracy is the ratio of the number of blocks hit during a time window to the number of blocks allocated to the VM. Here the workload is proj_4 from SNIA IOTTA Repository [24]. Observe from the figure that the proposed scheme reflecting the cache usage of entire window allows consistently better accuracy than S-CAVE. Notice that the average accuracy with the proposed scheme is consistently higher than that with S-CAVE, while it is much more stable.

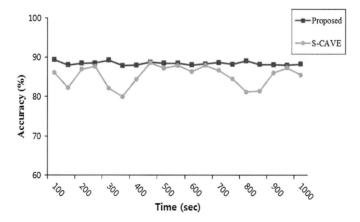

Fig. 5 The comparison of accuracies of the schemes

4.2 Cache Allocation

The evaluation on cache utilization with multiple VMs running with shared SSD cache is presented here. Figure 6 shows that the proposed scheme considerably reduces the amount of SSD cache space allocated to the VMs. It can be deemed that the proposed scheme uses SSD cache space more efficiently than the other scheme. Figure 7 compares hit ratio of the schemes. Both the proposed scheme and S-CAVE display similar hit ratio.

Two additional real-world traces are employed to evaluate the proposed scheme, which are MSR Cambridge trace from SNIA IOTTA Repository [24] and UMass Trace Repository [25] from a search engine. The traces represent a variety of workloads, hm_1(hardware monitoring) and Websearch. They generate I/O accesses at the storage disk tier and account for SSD cache as well as application caching effect.

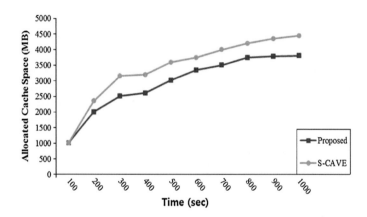

Fig. 6 The comparison of the amount cache space allocated to each VM

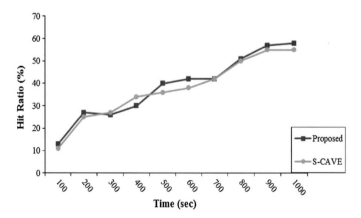

Fig. 7 The comparison of hit ratio of different schemes

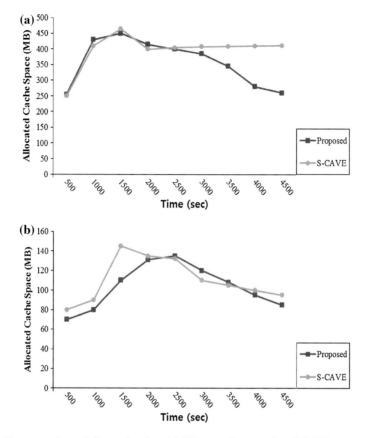

Fig. 8 The comparison of allocated cache with different schemes, **a** hm_1, **b** Websearch

Figure 8 shows that the proposed scheme slightly reduces the amount of SSD cache space compared to S-CAVE.

5 Conclusion

In this paper a hypervisor-based SSD caching scheme has been presented, which effectively manages the SSD storage cache in the VM environment by properly collecting and exploiting the runtime status of the VMs. A new metric was employed to accurately identify the demand on the cache space of each VM. The ratio of available cache space has also been accounted for dynamic adjustment of cache allocation among the VMs. Computer simulation validates the effectiveness the proposed scheme in achieving higher accuracy in the estimation of cache space for the VMs compared to the existing scheme. This allows the proposed scheme to display comparable hit ratio with less amount of SSD cache.

As future study, the proposed scheme will be enhanced for effective allocation with clustered SSD cache. Thereby, it will be able to provide flexible allocation of cache space and hardware resource in large-scale cloud environment. In addition to SSD cache, other shared resources such as CPU, memory, disk, network will be investigated for efficient resource management in the VM environment.

Acknowledgments This research was supported by Basic Science Research Program through the National Research Foundation of Korea (NRF) funded by the Ministry of Education, Science and Technology (2012R1A12040257 and 2014R1A1A2060398), the second Brain Korea 21 PLUS project, MSIP(Ministry of Science, ICT & Future Planning), Korea in the ICT R&D Program 2014 (1391105003), and Samsung Electronics (S-2014-0700-000).

References

1. Canim, M., Mihaila, G., et al.: SSD bufferpool extensions for database systems. In: Proceedings of the VLDB, pp. 1435–1446 (2010)
2. Luo, T., Lee, R., Mesnier, M.P., Chen, F., Zhang, X.: hStorage-DB: Heterogeneity aware data management to exploit the full capability of hybrid storage systems. In: PVLDB, pp. 1076–1087 (2012)
3. Qureshi, M.K., Patt, Y.N.: Utility-based cache partitioning: a low-overhead, high-performance, runtime mechanism to partition shared caches. In: MICRO (2006)
4. Smith, A.: Disk cache-miss ratio analysis and design considerations. ACM Trans. Comput. Syst. **3**, 161–203 (1985)
5. Kgil, T., Roberts, D., Mudge, T.: Improving NAND flash based disk caches. In: ISCA (2008)
6. Luo, T., et al.: S-CAVE: effective SSD caching to improve virtual machine storage performance. In: Proceedings of the 22nd International Conference on Parallel Architectures and Compilation Techniques, pp. 103–112 (2013)
7. Pritchett, T., Thottethodi, M.: Sievestore: a highly-selective, ensemble-level disk cache for cost-performance. In: Proceedings of the 37th International Symposium on Computer Architecture (ISCA 2010), pp. 163–174 (2010)
8. Stewart, C., Chakrabarti, A., Griffith, R.: Zoolander: efficiently meeting very strict, low-latency SLOs. In: Proceedings of the 10th International Conference on Autonomic Computing (ICAC), pp. 265–277 (2013)
9. Timothy, Z., Anshul, G., et al.: Saving cash by using less cache. In: Proceedings of the 4th USENIX conference on Hot Topics in Cloud Ccomputing (2012)
10. Jinho, H., Wei, Z., et al.: UniCache: Hypervisor managed data storage in RAM and flash. In: 2014 IEEE 7th International Conference, pp. 216–223 (2014)
11. Narayanan, D., Thereska, E., Donnelly, A., Elnikety, S., Rowstron, A.: Migrating server storage to ssds: analysis of tradeoffs. In: Proceedings of the 4th ACM European Conference on Computer Systems, pp. 145–158. ACM, New York (2009)
12. Anchev, N., et al.: Optimal cache replacement policy for matrix multiplication. In: ICT Innovations 2012, pp. 71–80. Springer, Berlin (2012)
13. Park, S., Jung, D., Kang, J., Kim, J., Lee, J., CFLRU: a replacement algorithm for flash memory. In: Proceedings of International Conference on Compilers, pp. 234–241 (2006)
14. Shim, H., Seo, B., Kim, J., Maeng, S.: An adaptive partitioning scheme for DRAM-based cache in solid state drives. In: Proceedings of the IEEE 26th Symposium on Mass Storage Systems and Technologies (2010)
15. Jinjiang L., Yihua L., et al.: An efficient schema for cloud systems based on SSD cache technology. Math. Probl. Eng. **2013**, 9 (2013) Article ID 109781

16. Jennings, B., Stadler, R.: Resource management in clouds: survey and research challenges. J. Netw. Syst. Manag. **23**(3), 731–737 (2015)
17. Gulati, A., Shanmuganathan, G., Zhang, X., Varman, P.J.: Demand based hierarchical QoS using storage resource pools. In: Proceedings of 2012 USENIX Annual Technical Conference (ATC 2012), USENIX (2012)
18. Liao, X., Jin, H., Yu, J., Li, D.: A performance optimization mechanism for SSD in virtualized environment. Comput. J. **56**, 992–1000 (2013)
19. Jinjiang, L., et al.: An efficient schema for cloud systems based on SSD cache technology. Math. Probl. Eng. **2013**, 9 (2013) Article ID 109781
20. Krishnaveni, N., Sivakumar, G.: Survey on dynamic resource allocation strategy in cloud computing environment. Int. J. Comput. Appl. Technol. Res. (IJCATR) **2**(6), 731–737 (2013)
21. Ahn, J., Kim, C., Choi, Y.R., Huh, J.: Dynamic virtual machine scheduling in clouds for architectural shared resources. In: Proceedings of 4th USENIX Workshop on Hot Topics in Cloud Computing (HotCloud 2012) (2012)
22. Jiang, S., Chen, F., Zhang, X.: CLOCK-Pro: an effective improvement of the CLOCK replacement. In: Proceedings of the USENIX '05 (April 2005)
23. Janapsatya, A., Ignjatović, A., Peddersen, J., Parameswaran, S.: Dueling clock: adaptive cache replacement policy based on the clock algorithm. In: Proceedings of the Conference on Design, Automation and Test in Europe, DATE 2010, pp. 920–925 (2010)
24. SNIA IOTTA Repository. http://iotta.snia.org/ (2011)
25. UMass Trace Repository. http://traces.cs.umass.edu/index.php/ (2007)

Emotional Scene Retrieval from Lifelog Videos Using Evolutionary Feature Creation

Hiroki Nomiya and Teruhisa Hochin

Abstract For the purpose of promoting the utilization of a large amount of lifelog videos, an emotional scene retrieval framework is proposed. It detects emotional scenes on the basis of facial expression recognition assuming that a kind of emotion will be aroused with a certain facial expression in an important scene which is likely to be a target of the retrieval. The emotional scene retrieval has a critical issue that it is quite hard to accurately and efficiently detect the emotional scenes because of the difficulty in discriminating spontaneous facial expressions. One of the most effective way to enhance the performance of the retrieval is to select discriminative facial features used for the facial expression recognition. It is, however, not easy to manually select good facial features because very subtle and complex movements of several facial parts will be observed in the appearance of a facial expression. We thus propose a method to automatically generate discriminative facial features on the basis of genetic programming. It produces discriminative facial features by combining a number of points on some salient facial parts using various arithmetic operators. The proposed method is evaluated through an emotional scene detection experiment using a lifelog video dataset containing spontaneous facial expressions.

Keywords Lifelog · Video retrieval · Facial expression recognition · Genetic programming

1 Introduction

Lifelog has recently attracted attention [1, 2]. It aims to record one's entire life as various types of data such as texts, images, and videos. In particular, lifelog videos [3] become more and more popular because of the emergence of easy-to-use video

H. Nomiya (✉) · T. Hochin
Kyoto Institute of Technology, Goshokaido-cho, Matsugasaki, Sakyo-ku, Kyoto
606-8585, Japan
e-mail: nomiya@kit.ac.jp

T. Hochin
e-mail: hochin@kit.ac.jp

© Springer International Publishing Switzerland 2016
R. Lee (ed.), *Software Engineering, Artificial Intelligence, Networking
and Parallel/Distributed Computing 2015*, Studies in Computational Intelligence 612,
DOI 10.1007/978-3-319-23509-7_5

recording devices. However, lifelog videos have a crucial problem that it is difficult to accurately and efficiently retrieve useful scenes from large-scale video databases. As a result, valuable lifelog videos often remain unused.

For the purpose of promoting the utilization of lifelog videos, we propose an impressive scene retrieval method for lifelog videos. The proposed method retrieves emotional scenes such as the scenes in which a person in the video is smiling, considering that a certain impressive event could happen in most of emotional scenes [4]. The emotional scenes are detected on the basis of facial expression recognition because most of emotions can be reflected in the facial expressions.

Facial expression recognition has been applied to video-scene detection [5, 6]. Most of the facial expression recognition techniques manually define their own facial features and discriminate facial expression using them. The facial feature is one of the core elements in the facial expression recognition and dominates the recognition performance. It is, however, not easy to manually select good facial features because a variety of very subtle and complex movements of several facial parts will be observed in the appearance of a facial expression.

In order to facilitate the creation of discriminative facial features, we introduce genetic programming [7]. Genetic programming is widely used to automatically generate computer programs that perform some kind of user-defined tasks. Since a program is represented by a tree structure, our facial feature can be represented as a form of a tree by combining a number of points on some salient facial parts (as the terminal nodes) using various arithmetic operators (as the non-terminal nodes). We attempt to create discriminative facial features by defining an effective fitness function to evaluate the discrimination ability of the facial features in the genetic operations.

For the detection of the emotional scenes, we introduce an efficient facial expression recognition method using the facial features and an emotional scene detection method on the basis of the recognition result for each frame image in a video.

We show the effectiveness of the the proposed method through an emotional scene retrieval experiment using several lifelog videos including spontaneous facial expressions.

The remainder of this paper is organized as follows. Section 2 presents related works. Section 3 illustrates the facial features. Section 4 explains the creation of facial features. Section 5 shows the facial expression recognition method. Section 6 elaborates the emotional scene detection method. Section 7 describes an emotional scene detection experiment. Finally, Sect. 8 concludes this study.

2 Related Works

The performance of the emotional scene detection is largely dependent on the facial expression recognition. While there are various approaches to recognize facial expressions [8], our approach is based on the geometric features taking into consideration the tradeoff between the accuracy and efficiency. The geometric features describe the shape and locations of salient facial components such as eyebrows, eyes, and a mouth.

For example, 3D models of the faces are used as the geometric features in order to accurately recognize facial expressions [9]. Although they can be very precise and effective for accurate recognition, it will be difficult to utilize the 3D facial features within reasonable cost considering that a lifelog video database can be very large.

By using 2D models, the geometric features can be more concise. They are defined as the positional relationships of the feature points on a face such as the distance between two points [4, 10]. Most of the facial features are manually determined by combining some points. It is, however, unclear whether they are sufficient to accurately distinguish various facial expressions.

Our method has the possibility to acquire more powerful facial features since the genetic programming selects useful facial features through the evaluations and genetic operations to a very large number of possible facial features.

3 Facial Feature

3.1 Facial Feature Points

We introduce *facial feature points*, which are the points on salient facial parts, in order to recognize facial expressions. They are introduced with the intention to achieve the balance between the conciseness and discrimination ability of the facial features. The facial feature points are obtained by an application software called FaceSDK 4.0 [11].

We utilize 32 facial feature points on the right half of a face considering the symmetric feature of the facial parts. The facial features are shown in Fig. 1. The white squares represent the points detected by FaceSDK and the points denoted by p_1 to p_{32} are the facial feature points used in the proposed method.

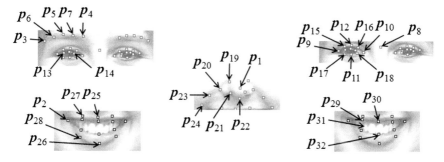

Fig. 1 Facial feature points

3.2 Structure of Facial Feature

The facial features are defined on the basis of the facial feature points. Taking the tradeoff between the accuracy and efficiency of the facial features into consideration, the facial features are concisely represented by the positional relationship of several facial feature points.

However, a single positional relationship such as the distance between two points is too simple to accurately distinguish subtle differences of the facial expressions. In order to make the facial features more discriminative, we introduce genetic programming [7] into the creation of the facial features. In the proposed method, a facial feature corresponds to a program and can be represented as a tree.

The tree representation of the facial feature consists of two types of nodes called terminal nodes and non-terminal nodes. The terminal (non-terminal, respectively) nodes are the leaf (non-leaf) nodes in the tree. A tree represents a certain facial feature as a whole. A non-terminal node can be regarded as an operator and the terminal nodes are the operands for the operator. Tables 1 and 2 describe the terminal and non-terminal nodes used in the proposed method, respectively.

The terminal nodes can be either a vector or a scalar. The vector-type terminal nodes are the vector representations of the facial feature points on the two-dimensional Euclidean space. The scalar-type terminal nodes are the integer constants used, for example, for the coefficients of a certain term.

Table 1 Terminal nodes

Type	Number of nodes	Nodes
Vector	32	$p_1, \ldots, p_{32}$
Scalar	4	$1, 2, 3, 4$

Table 2 Non-terminal nodes

Node	#Arguments	Operation	
		Vector	Scalar
N_1	2	$v_1 + v_2$	$s_1 + s_2$
N_2	2	$v_1 - v_2$	$s_1 - s_2$
N_3	2	$v_1 \cdot v_2$	$s_1 \times s_2$
N_4	2	$v_1 \times v_2$	s_1/s_2
N_5	1	$\sin \theta(v_1)$	$\sqrt{|s_1|}$
N_6	1	$\cos \theta(v_1)$	s_1^2
N_7	1	$\tan \theta(v_1)$	$\log(|s_1| + 1)$
N_8	1	$|v_1|$	$|s_1|$
N_9	2	$|v_1 - v_2|$	$\min\{s_1, s_2\}$
N_{10}	2	$|\theta(v_1) - \theta(v_2)|$	$\max\{s_1, s_2\}$

[a] v_1 and v_2 are the vector-type arguments.
[b] s_1 and s_2 are the scalar-type arguments.
[c] $\theta(v_i)$ is the angle of v_i in the polar coordinate system.

Fig. 2 Example of facial features

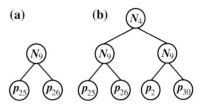

We define ten types of non-terminal nodes. The operations of a non-terminal node are separately defined for the vector arguments and the scalar arguments. For example, N_1 yields the vector of the sum of v_1 and v_2 for the vector-type arguments v_1 and v_2. For the scalar-type arguments s_1 and s_2, it yields the scalar corresponding to $s_1 + s_2$. In the case that one of the arguments is a vector and the other is a scalar, the operation for N_1 is not defined. We thus introduce a restriction in the genetic programming so that the tree including such arguments should not be generated.

Figure 2a describes a simple example of the facial feature. N_9 is regarded as the operation $|v_1 - v_2|$ since both of the arguments (i.e., p_{25} and p_{26}) are the vector-type ones. This facial feature value is, therefore, represented as $|p_{25} - p_{26}|$, which means the Euclidean distance between the two facial feature points p_{25} and p_{26}.

Figure 2b shows a little more complex example. The left and right subtrees yield the scalar values $|p_{25} - p_{26}|$ and $|p_2 - p_{30}|$, respectively. N_4 receives two scalar-type arguments and produces the facial feature value $|p_{25} - p_{26}|/|p_2 - p_{30}|$, which indicates the vertical-to-horizontal ratio of the mouth.

4 Creation of Facial Features

It is very difficult to manually create useful facial features by combining the facial feature points because there are infinite possible combinations of the facial feature points and the operations for them. We thus introduce genetic programming [7] into the creation process of the facial features to make it easy to discover useful combinations.

The algorithm to create the facial features is shown in Algorithm 1. The proposed feature creation algorithm is based on a standard genetic programming technique including some commonly-used genetic operations such as crossover and mutation [7]. The key point of the proposed feature creation algorithm is the fitness function used to measure the goodness of each individual in the population.

The fitness of an individual is determined through the two-step fitness computation. In the first step, the tentative fitness value is computed for each individual on the basis of the ratio of the between-class and within-class variances since it is generally proportional to the discrimination performance. Additionally, as shown in Eq. (1), the size of an individual is taken into consideration in order to prevent the individual from being redundant. Some of the individuals cannot compute the feature

---

Algorithm 1 Creation of facial features.

Input:
- Training examples $\{(x_1, y_1), \ldots, (x_n, y_n)\}$ where n is the number of examples
  x_i : A set of 32 facial feature points
  y_i : Class label representing the facial expression appeared in the example
- Number of facial expressions C ($y_i \in \{1, \ldots, C\}$)
- Number of individuals N in the population
- Number of generations G
- Number of output facial features D
- Parameters α and M
- Several parameters for the genetic programming (explained in detail in the experimental section)

Procedure:

1: Create the first (initial) population P_1.

2: $g \leftarrow 1$.

3: For each individual $I \in P_g$, compute the tentative fitness value $\tilde{F}$ given by Eq. (1).

$$\tilde{F}(I) = \begin{cases} \frac{VR(I)}{1+\alpha N(I)} & \text{if } I \text{ is } valid \\ 0 & \text{otherwise} \end{cases} \tag{1}$$

where, $N(I)$ is the size of I (i.e., the number of nodes in I), and α is the parameter corresponds to the weight for the size of the individual. $VR(I)$ is the ratio of the between-class variance to the within-class variance given by Eq. (2).

$$VR(I) = \frac{\sum_{i=1}^{C} \frac{n_i}{n} (\mu_i - \mu)^2}{\frac{1}{n} \sum_{i=1}^{C} \sum_{j=1}^{n_i} (f(x_j^i) - \mu_i)^2} \tag{2}$$

where, $f(x)$ is the facial feature value computed from the tree of I and the facial feature points of x. x_j^i is the jth training example whose class label is i (i.e., $y = i$). n_i is the number of training examples having the class label i. μ_i and μ are obtained from Eq. (3):

$$\mu_i = \frac{1}{n_i} \sum_{j=1}^{n_i} f(x_j^i), \ \mu = \frac{1}{n} \sum_{i=1}^{n} f(x_i) \tag{3}$$

4: Compute the (final) fitness value F using Eq. (4).

$$F(I) = \begin{cases} (1 - |K(I)|)\tilde{F}(I) & \text{if } I \text{ is } valid \\ 0 & \text{otherwise} \end{cases} \tag{4}$$

where $K(I)$ is the penalty term obtained on the basis of the correlation between I and the other individuals having large value of $\tilde{F}$. For the individual I_r having the rth largest value of $\tilde{F}$, $K(I_r)$ is given by Eq. (5).

$$K(I_r) = \begin{cases} 0 & r = 1 \\ \max\limits_{i=1,\ldots,r-1} Cor(I_r, I_i) & 2 \leq r \leq M+1 \\ \max\limits_{i=1,\ldots,M} Cor(I_r, I_i) & r > M+1 \end{cases} \tag{5}$$

where $Cor(I_r, I_i)$ is Pearson's correlation coefficient between I_r and I_i.

5: $g \leftarrow g + 1$.

6: if $g > G$ then proceed to Step 7. Otherwise, create the gth population P_g using F and return to Step 3.

7: Finish the procedure and output D facial features from P_G having the first-to-Dth highest (final) fitness values.

---

values because of the type error of the arguments (for example, one of the argument is scalar-type while the other one is vector-type). Such individuals are called *invalid* and their tentative fitness values are set to 0. On the other hand, the individuals which are able to compute feature values are called *valid*.

It is meaningless to simply choose the individuals having high tentative fitness values because they tend to have very similar tree structures. Acquiring sufficient discrimination ability needs the diversity of the facial features as well as the higher (tentative) fitness values. In the second step, therefore, the final fitness values are determined using the correlation between the individuals. As shown in Eq. (4), the final fitness value is computed as the product of the tentative fitness value and the penalty term on the basis of the correlation. For example, the penalty term of the individual having rth highest tentative fitness value is obtained based on the maximum value of the Pearson's correlation coefficients between the individual and the individuals having the first-to-$(r - 1)$th highest tentative fitness values. In order to reduce the computational cost, the correlation coefficients are computed only for the individuals having the first-to-Mth highest tentative fitness values when $r > M + 1$. The parameter M should be experimentally determined.

At the end of the final generation, D individuals having the first-to-Dth highest (final) fitness values are output as the facial features. The parameter D should also be experimentally determined.

5 Facial Expression Recognition

As a result of the aforementioned feature creation, D facial features are obtained. Each facial feature has the tree structure which receives several facial feature points represented as vectors and outputs a (scalar) facial feature value. A D-dimensional feature vector can therefore be defined.

The feature vector $V(x)$ for a certain set of facial feature points x is described as $V(x) = (f_1(x), \ldots, f_D(x))$, where f_i is the value of the ith facial feature obtained through the feature creation. An example is thus described as $(V(x), y)$ where y is the class label.

The problem of the discrimination of the facial expression is represented as the problem of the determination of the class labels of given examples. This can be solved by using a certain machine learning technique. In the proposed method, we make use of Support Vector Machine (SVM) [12] which is widely used to predict the class labels of given feature vectors.

6 Emotional Scene Detection

The emotional scenes are detected from a video according to the predicted class label for each frame image. The emotional scenes with a certain facial expression are determined by using the frame images having the corresponding class labels.

At the first step of the emotional scene detection, each frame image having the corresponding class label is regarded as a single emotional scene. Then, neighboring emotional scenes are integrated into a single emotional scene. The integration process is repeated until no more emotional scenes can be integrated. The resulting scenes are output as the emotional scenes of the facial expression.

The algorithm to detect the emotional scenes is shown in Algorithm 2. Since the emotional scene detection algorithm can find the emotional scenes for a single facial expression, it is required to perform the emotional scene detection C times when there are C kinds of facial expressions in a video.

7 Experiment

7.1 Experimental Settings

7.1.1 Lifelog Videos

We prepared six lifelog videos by six subjects termed Subject A, B, C, D, E, and F. All the subjects are male university students.

The lifelog videos contain the scenes of playing cards recorded by web cameras. A single web camera recorded a single subject so that the subject's frontal face was recorded. This experimental setting is due to the limitation of FaceSDK that it can detect the facial feature points of a single frontal face. While card games are suitable for stably recording frontal faces, a player of most of card games tries to keep a poker face. We thus chose the card games such as Hearts in which the players could clearly express the emotion.

The lengths of the videos vary from 10.0 min to 13.3 min. The average length is 11.6 min. The size of each video is 640 × 480 pixels and the frame rate is 30 frames per second. Considering the high frame rate, we selected frames from each video after every 10 frames in order to reduce the computational cost. Consequently, the number of frame images in a video is 2088 on average.

Apart from the web cameras, a video camera was used to record the scenes of playing cards including all subjects at the same time. The videos were recorded in order to show the results of the emotional scene retrieval to the users while they were not used in this experiment. Note that the videos recorded are not shown in this paper because of privacy reasons.

Algorithm 2 Emotional scene detection.

Notations:

- E_i^c: The ith emotional scene in which the facial expression c appears.
- $first(E_i^c)$: Frame number of the beginning frame in E_i^c.
- $last(E_i^c)$: Frame number of the ending frame in E_i^c.
- $length(E_i^c)$: Length of E_i^c. It is equivalent to $last(E_i^c) - first(E_i^c) - 1$.
- $\#int(E_i^c)$: Number of emotional scenes integrated into E_i^c.
- $\#nonemo(E_i^c)$: Number of nonemotional frames in E_i^c.
  Note that a nonemotional frame means that the facial
  expression appears in that frame is different from c.
- $dist(E_i^c, E_j^c)$: The distance between E_i^c and E_j^c ($i < j$).
  It is equivalent to $first(E_j^c) - last(E_i^c) - 1$.

Initialize:

For each frame image having the class label c, perform the following initialization according to Eq. (6):

$$first(E_i^c) = last(E_i^c) = c_i, \ \#int(E_i^c) = 0,$$

$$\#nonemo(E_i^c) = 0, \ length(E_i^c) = 1, \ (1 \le i \le M_c) \qquad (6)$$

where, c_i is the frame number of the ith emotional frame in the video. M_c is the number of emotional frames. An emotional frame is the frame having the class label c. That is, each emotional scene consists of a single emotional frame.

Procedure:

1: Find i^* in accordance with Eq. (7):

$$i^* = \operatorname*{argmin}_i \, dist(E_i^c, E_{i+1}^c)$$

$$s.t. \ dist(E_i^c, E_{i+1}^c) \le \frac{length(E_i^c) - \#nonemo(E_i^c)}{\#int(E_i^c) + 1}$$

$$\wedge \, dist(E_i^c, E_{i+1}^c) \le \frac{length(E_{i+1}^c) - \#nonemo(E_{i+1}^c)}{\#int(E_{i+1}^c) + 1} \qquad (7)$$

2: If there is no i^* that satisfies Eq. (7), finish the procedure and output current emotional scenes. Otherwise, proceed to step 3.

3: Integrate $E_{i^*+1}^c$ into $E_{i^*}^c$ by updating $E_{i^*}^c$ as follows:

$$last(E_{i^*}^c) \leftarrow last(E_{i^*+1}^c), \ \#int(E_{i^*}^c) \leftarrow \#int(E_{i^*}^c) + 1,$$

$$\#nonemo(E_{i^*}^c) \leftarrow \#nonemo(E_{i^*}^c) + first(E_{i^*+1}^c)$$
$$-last(E_{i^*}^c) - 1$$

Note that $length(E_{i^*}^c)$ is also updated due to the update of $last(E_{i^*}^c)$.

4: Delete $E_{i^*+1}^c$ and renumber the subscripts of E_i^c so that the emotional scenes become $E_i^c, \ldots, E_{M_c-1}^c$.

5: $M_c \leftarrow M_c - 1$ and return to step 1.

The facial expressions observed in most of the emotional scenes in the lifelog videos were smiles. Thus, we set the value of C (described in Algorithm 1) to 2 intending to detect the emotional scenes with smiles, that is, to discriminate smiles and other facial expressions. The ratio of the emotional frames to all the frames varies from 16.6 to 29.6 %. A subject is smiling in 24.6 % of the frames in the video on average.

A two-fold cross validation was used in this experiment by dividing each video into the former and latter parts of the same lengths. One of the part was used for the training and the other part was used for the test.

7.1.2 Facial Feature Creation

The genetic programming module was implemented using a C++ library called GPC++ [13]. The parameters for the genetic programming were experimentally determined as shown in Table 3.

The parameters α and M in Algorithm 1 were set to 0.001 and 100 respectively according to the result of the preliminary experiment. The experimental result is not described here because of the space limitation but these parameters lead to relatively good performance. The number of facial features D was set to 1 to 10 since it seemed to have greater influence on the performance compared with the other parameters.

7.1.3 Facial Expression Recognition

SVM is used for the discrimination of the facial expression. We implemented it using LIBSVM [14] and used linear C-SVC.

Table 3 Parameters for genetic programming

Parameter	Value
Population size (Number of individuals N)	10000
Number of generations G	100
Selection method	Roulette wheel selection
Probability of crossover	0.85
Probability of mutation	0.10
Probability of copying existing individual	0.05
Population initialization method	Ramped-half-and-half
Maximum tree depth for crossover	20
Maximum tree depth for mutation	10

7.2 Experimental Result

7.2.1 Emotional Scene Detection Accuracy

The recall, precision, and F-measure of the emotional scene detection are computed for the evaluation of the accuracy of the proposed method. The recall, precision, and F-measure are defined by Eqs. (8), (9) and (10), respectively.

$$recall = \frac{|T \cap \hat{T}|}{|T|} \tag{8}$$

$$precision = \frac{|T \cap \hat{T}|}{|\hat{T}|} \tag{9}$$

$$F-measure = \frac{2 \cdot recall \cdot precision}{recall + precision} \tag{10}$$

where, T is the correct set of emotional frames. One of the authors determined whether each frame was emotional or not prior to the experiment. $\hat{T}$ is the set of emotional frames detected by the proposed method.

The recall, precision, and F-measure of the emotional scene detection for each subject are shown in Figs. 3, 4 and 5, respectively. The average recall, precision, and F-measure on the cross validation for each value of D are described in these figures.

The value of D does not have significant effect on the accuracy except for Subject D. This indicates that a small number of facial features is sufficient for many people. It could be a merit of the proposed method because a smaller number of facial features leads to the smaller computational cost.

Fig. 3 Recall of emotional scene detection

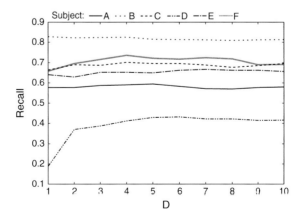

Fig. 4 Precision of
emotional scene detection

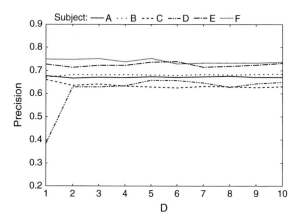

Fig. 5 F-measure of
emotional scene detection

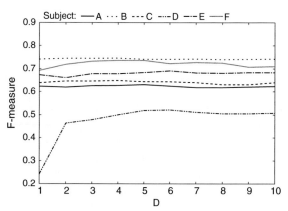

The accuracy of Subject D is lower than those of the other subjects. This means that it is relatively difficult to discriminate his facial expression. A very small number of facial features will not be sufficient in such a case.

The best accuracy (F-measure) is obtained when $D = 4$ for Subject B and C, $D = 5$ for Subject A and F, and $D = 6$ for Subject D and E. From this observation, six facial features seem to be sufficient since more than six facial features do not improve the accuracy and increase the computational cost.

7.2.2 Emotional Scene Detection Efficiency

For the evaluation of the efficiency, the average computational time for each subject when $D = 6$ is shown in Fig. 6.

A computer with a Xeon W3580 CPU (3.33GHz) and 16GB memory was used. Parallel processing was not used in this experiment. The tree structures of the facial features and the learned SVM model were stored in the external files in advance

Fig. 6 Computational time of emotional scene detection

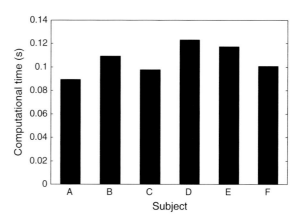

(at the learning phase). Note that the computational time to obtain the facial feature points is not included because this process is performed by an existing software application.

While the computational time is a little different from subject to subject due to the difference of the structure of the facial features, the emotional scene detection takes only about 0.1 seconds for every subject. This means that it will take less than one minute for the emotional scene detection from a video with 24-hour length, under the conditions that the frame rate of the video is reduced in advance, there is a memory space enough to store the entire video, and the learning phase is completed. Our method is fully practical since it is not difficult to satisfy these conditions. Because of this efficiency, the proposed method can be applied to large-scale video databases.

7.2.3 Generated Facial Features

The facial features generated are shown in this section. Due to the space limitation, only two examples are shown here.

One of the facial features generated for Subject A (Subject B, respectively) is described as f_A (f_B) in Eqs. (11), (12). These are the equational representations of the tree structures of the individuals. p_i is the vector representation of the ith facial feature point. $\theta(v)$ is the angle of the vector v in the polar coordinate system. Note that the redundant nodes (e.g., the node of absolute value (N_8) which is the parent of the node of square (N_6)) are eliminated for the better readability.

$$f_A = \sqrt[4]{|\theta(p_6 - p_{24}) - \theta(p_2)|} \tag{11}$$
$$f_B = \log(\{\theta(p_2) - \theta(p_{13} - p_{24} + p_{22})\}^2 + 1) \tag{12}$$

The facial feature points p_2 and p_{24} frequently appear in the facial features of all the subjects as well as A and B. p_2 is located on the corner of the mouth and p_{24} is

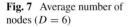
Fig. 7 Average number of nodes ($D = 6$)

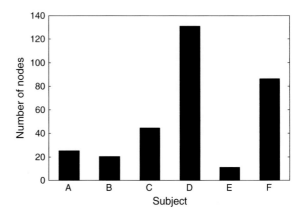

one of the points on the nasolabial folds. These points are considered to be important for the detection of smiles.

For each subject, the average number of nodes in the six individuals used as the facial features is shown in Figure 7. The number of nodes is very different from subject to subject. The individuals of Subject D have larger number of nodes compared with the other subjects. Considering that the emotional scene detection accuracy of Subject D is significantly lower than that of the other subjects, this result could stem from the difficulty of the discrimination of his facial expression.

The individuals of Subject F also have a large number of nodes in spite of its relatively high emotional scene detection accuracy. This could be attributed to the tendency that some of the individuals of Subject F contain a number of redundant nodes. Although the degradation of the efficiency of the emotional scene detection caused by the redundant nodes seems to be very small, it is necessary to improve the proposed method so that it can produce more concise individuals. Note that the computational time is not always proportional to the number of nodes because the computational complexity of each mathematical operation (of the corresponding non-terminal node) is not uniform.

8 Conclusion

An emotional scene retrieval method is proposed for the purpose of promoting the utilization of lifelog videos. The proposed method focuses on the creation of useful facial features and provides the methodology of generating the facial features effective for the discrimination of the facial expressions on the basis of the genetic programming. The experimental result shows that the proposed method can efficiently retrieve the emotional scenes from several lifelog videos.

In the experiment, we retrieved only the emotional scenes with smiles. Evaluating the retrieval performance for the scenes with the other emotions is thus included in the future works. In addition, discriminating various kinds of smiles such as a full smile and a wry smile is important for improving the usability of the emotional scene retrieval system.

The genetic programming has a variety of parameters as shown in Table 3. Some of them could have great influence on the performance of the emotional scene detection. Providing the efficient method to tuning the parameters is also included in the future work.

Acknowledgments This research is supported by Japan Society for the Promotion of Science, Grant-in-Aid for Young Scientists (B), 15K15993.

References

1. Aizawa, K., Hori, T., Kawasaki, S., Ishikawa, T.: Capture and efficient retrieval of life log. In: Proceedings of Pervasive 2004 Workshop on Memory and Sharing Experiences, pp. 15–20 (2004)
2. Gemmell, J., Bell, G., Luederand, R., Drucker, S., Wong, C.: MyLifeBits: fulfilling the memex vision. In: Proceedings of the 10th ACM International Conference on Multimedia, pp. 235–238 (2002)
3. Datchakorn, T., Yamasaki, T., Aizawa, K.: Practical experience recording and indexing of life log video. Proceedings of the 2nd ACM Workshop on Continuous Archival and Retrieval of Personal Experiences, pp. 61–66 (2005)
4. Nomiya, H., Morikuni, A., Hochin, T.: An unsupervised ensemble approach for emotional scene detection from lifelog videos. Softw. Eng. Artif. Intell. Netw. Parallel/Distrib. Comput. Stud. Comput. Intell. **569**, 145–159 (2015)
5. Datcu, D., Rothkrantz, L.: Facial expression recognition in still pictures and videos using active appearance models: a comparison approach. In: Proceedings of the 2007 International Conference on Computer Systems and Technologies, pp. 1–6 (2007)
6. Fanelli, G., Yao, A., Noel, P.-L., Gall, J., Gool, L.V.: Hough forest-based facial expression recognition from video sequences. In: Proceedings of the 11th European Conference on Trends and Topics in Computer Vision, pp. 195–206 (2010)
7. Koza, J.R.: Genetic programming: on the programming of computers by means of natural selection. The MIT Press, Cambridge (1992)
8. Tian, Y., Kanade, T., Cohn, J.F.: In: Li, S.Z., Jain, A.K. (eds.) Handbook of face recognition. Facial Expression Recognition. Springer, London (2011)
9. Soyel, H., Demirel, H.: Facial expression recognition using 3d facial feature distances. In: Proceedings of the 4th International Conference on Image Analysis and Recognition, pp. 831–838 (2007)
10. Hupont, I., Cerezo, E., Baldassarri, S.: Sensing facial emotion in a continuous 2D affective space. In: Proceedings of International Conference on Systems, Man, and Cybernetics, pp. 2045–2051 (2010)
11. Luxand Inc.: Luxand FaceSDK 4.0 (2015). http://www.luxand.com/facesdk (The latest version is 5.0.1)
12. Cortes, C., Vapnik, V.: Support-vector networks. Mach. Learn. **20**(3), 273–297 (1995)
13. Fraser, A., Weinbrenner, T.: GPC++ - Genetic Programming C++ Class Library, Version 0.5.2 (2015). http://www0.cs.ucl.ac.uk/staff/ucacbbl/ftp/weinbenner/gp.html
14. Chang, C.-C., Lin, C.-J.: LIBSVM: a library for support vector machines. ACM Transactions on Intelligent Systems and Technology, **2**(3, 27):1–27 (2011)

On Solving the Container Problem
in a Hypercube with Bit Constraint

Antoine Bossard and Keiichi Kaneko

Abstract As shown in the TOP500 list, hypercubes are popular as interconnection networks of massively parallel systems. This popularity comes mainly from the simplicity and ease of implementation of this topology. To avoid bottleneck situations, communication algorithms and routing in general is a critical topic for these high-performance systems. It has been shown that disjoint paths routing is a very desirable property for these communication algorithms. Effectively, disjoint paths ensure the absence of infamous parallel processing issues such as deadlocks, livelocks and starvations. In this paper, we propose a routing algorithm selecting in a hypercube internally node-disjoint paths between any two nodes, and such that the selected paths all satisfy a given bit constraint. This bit constraint mechanism enables the selection of multiple sets of disjoint paths between several node pairs each satisfying a distinct bit constraint, something impossible with conventional routing algorithms. The simultaneous selection of disjoint paths between different node pairs offers even better communication performance and system dependability. The correctness of the proposed algorithm is formally established and empirical evaluation is conducted to inspect the algorithm practical behaviour.

Keywords Supercomputer · Parallel system · Network · Cube · Routing algorithm · Disjoint paths · Dependable system · Node-to-node

A. Bossard (✉)
Graduate School of Science, Kanagawa University, 2946 Tsuchiya, Hiratsuka,
Kanagawa 259-1293, Japan
e-mail: abossard@kanagawa-u.ac.jp

Keiichi Kaneko
Graduate School of Engineering, Tokyo University of Agriculture and Technology,
Nakacho 2-24-16, Koganei, Tokyo 184-8588, Japan
e-mail: k1kaneko@cc.tuat.ac.jp

© Springer International Publishing Switzerland 2016
R. Lee (ed.), *Software Engineering, Artificial Intelligence, Networking
and Parallel/Distributed Computing 2015*, Studies in Computational Intelligence 612,
DOI 10.1007/978-3-319-23509-7_6

77

1 Introduction

Due to their simplicity and thus ease of hardware and software implementation, hypercubes [1] are popular as interconnection networks of massively parallel systems. Such machines featuring a decades long history [2], the supercomputers NASA Pleiades and NOAA Zeus [3] are two very recent examples. In addition, it is worth noticing that hypercubes are also very popular as seed or sub-network of advanced interconnection network topologies, and especially those for hierarchical interconnection networks (HINs), such as dual-cubes [4], metacubes [5], hierarchical hypercubes [6] and hierarchical cubic networks [7].

For these reasons, hypercube routing is an actively researched topic. Several hypercube routing algorithms have been proposed in the literature: optimal node-to-node disjoint paths routing algorithm [8, 9], node-to-set disjoint paths routing algorithm [10], set-to-set disjoint paths routing algorithm [11] and k-pairwise disjoint paths routing algorithm [12] are some examples. These conventional approaches do not allow for enforcing a constraint on the nodes selected in the paths. Yet, routing with constraint offers interesting properties, enables new applications, and induces better performance and system dependability. Effectively, by enforcing such a restriction on nodes, we can easily achieve disjoint paths routing between several nodes pairs, each satisfying a distinct node constraint. This has been first discussed in [13] and is a convenient way to obtain a path signature as defined in [14].

Disjoint paths routing is a critical property for communication algorithms in parallel systems. Indeed, routing according to mutually disjoint paths guarantees that infamous resource allocation problems in parallel systems, such as deadlocks, livelocks and starvations, shall never occur. In addition, the simultaneous path selection and thus data communication is obviously important for performance matters: parallel communication allows for a more efficient usage of the network, with implications going as far as Green IT [15]. Another important aspect of disjoint paths routing is that it dramatically increases system dependability. Effectively, considering the huge number of computing nodes included in modern supercomputers (e.g. 705,024 in the Fujitsu K [3]), it is unavoidable that faults (i.e. broken nodes) will occur [16]. By selecting mutually node-disjoint paths, the impact of one such faulty node is severely limited: one fault can jeopardise at most one path due to the mutual disjointness of the selected paths.

Now, we describe in this paper a routing algorithm solving the container problem with bit constraint in a hypercube. The container problem is also known as the node-to-node disjoint paths routing problem and is about finding a set of mutually (internally) node-disjoint paths between any pair of nodes [17–19]. The possibilities offered by this algorithm further increase the advantages induced by routing with bit constraint. Concretely, by considering several pairs of nodes with a distinct bit constraint for each pair, it is possible to easily find *at the same time* several sets of disjoint paths between these node pairs, something which is impossible with conventional algorithms, even disjoint paths routing algorithms. This enhanced capability of selecting disjoint paths that are available at the same time for routing induces improved

Fig. 1 Several sets of node-to-node disjoint paths, enabling high performance network communications

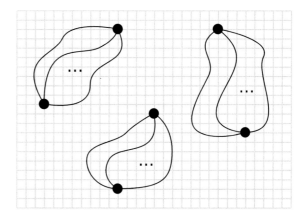

communication performance and increased system dependability. An illustration is given in Fig. 1.

The rest of this paper is organised as follows. We recall in Sect. 2 notations and definitions used throughout the paper. In Sect. 3.1, we describe the hypercube node-to-node disjoint paths routing algorithm with bit constraint, additionally providing pseudo-code for this algorithm. Next, we show the algorithm correctness and establish the algorithm complexities in Sect. 3.2. An example of the algorithm execution trace is given in Sect. 3.3. In Sect. 4, we conduct an empirical evaluation of the proposed algorithm in order to inspect its practical behaviour and compare it with the theoretical results obtained in the previous sections. Finally, this paper is concluded in Sect. 5.

2 Preliminaries

In this section, we recall several definitions and notations used hereinafter. Also, additional notations are introduced, and previous results given.

Definition 1 An n-dimensional hypercube, denoted by Q_n, consists of 2^n nodes, each having a unique n-bit address. Two nodes u and v of a hypercube are adjacent if and only if their Hamming distance $H(u, v)$ is equal to one.

Regarding topological properties of a hypercube, we have that a Q_n is symmetric and of connectivity, degree and diameter n [1]. Additionally, it is important to note that a Q_n has a recursive structure. Effectively, for any dimension δ $(0 \leq \delta \leq n-1)$, a Q_n consists of two $(n-1)$-dimensional hypercubes Q_{n-1}^0 and Q_{n-1}^1, called subcubes, and defined as follows. The subcube Q_{n-1}^0 (resp. Q_{n-1}^1) is induced by the set of nodes of Q_n whose δth bits are set to 0 (resp. 1). When considering the subcubes of a hypercube, we talk of *hypercube reduction*. A 4-dimensional hypercube Q_4 with its two subcubes Q_3^0 and Q_3^1 highlighted is illustrated in Fig. 2.

Fig. 2 A 4-dimensional hypercube Q_4 with its two subcubes Q_3^0 and Q_3^1 induced by $\delta = 0$

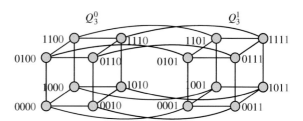

Now, let us assume that the address of any node of a Q_n, that is an n-bit address, can be stored in a fixed number of machine words, thus enabling constant time node comparison, most significant bit (MSB) detection as well as Hamming distance and bit weight (as defined in Definition 2) calculations.

Definition 2 For a binary n-bit sequence $b = b_{n-1} \ldots b_1 b_0$, $b_i \in \{0, 1\}$, $0 \leq i \leq n - 1$, the *bit weight* of b, denoted by $w(b)$, is the number of bits of b that are set to 1.

In addition, we adopt the following conventions: logarithms mentioned in this paper are in base two; the MSB of bit sequence is the leftmost bit. Furthermore, we use the following notations to denote the binary bitwise operations: the bitwise AND is denoted by &, the binary negation is denoted by ¬, and the bitwise exclusive-OR is denoted by ⊕.

Definition 3 A k-constraint is a k-tuple of distinct natural numbers $(i_1, i_2, \ldots, i_k)$.

In this paper, a constraint is applied to the bit weight of a hypercube node. We focus on 2-constraints and simply speak of *bit constraints*, which are denoted by pairs of natural numbers (i, j). As we consider routing inside hypercubes where we recall that adjacent nodes have one single bit different, it is easy to understand that bit constraints considered all have the form $(i, i + 1)$. And if we were to consider k-constraints on hypercubes, those bit constraints would have the form $(i, i + 1, \ldots, i + \beta)$ with $i + \beta \leq n$.

Definition 4 In a Q_n, for $i \in \mathbb{N}$ and $0 \leq i \leq n - 1$, a node u satisfies the constraint $\gamma_i = (i, i + 1)$ if and only if $w(u) = i$ or $w(u) = i + 1$ holds.

As an example, let us consider a Q_3 and the bit constraint $\gamma_1 = (1, 2)$. Then, the three nodes 010, 110 and 100 all satisfy γ_1 whereas the node 111 does not.

The remaining definitions and notations to be recalled deal with paths. First, a path in a graph is an alternate sequence of nodes and edges; for a path p, we write $p : u_1, (u_1, u_2), u_2, \ldots, u_{k-1}, (u_{k-1}, u_k), u_k$, with (u_i, u_{i+1}) denoting the edge between the two distinct nodes u_i and u_{i+1}. For convenience, that same path p can also be written as $u_1 \rightarrow u_2 \rightarrow \ldots \rightarrow u_k$ and even more concisely as $u_1 \rightsquigarrow u_k$, the latter notation possibly bringing ambiguity regarding the nodes included in the path and thus a notation to be used with care. The length of a path corresponds to the number of its edges.

We say that two paths are mutually node-disjoint (we simply say *disjoint*) when they have no node in common. For convenience we consider a path as a set of nodes, and thus the two paths p_1 and p_2 are disjoint if and only if $p_1 \cap p_2 = \emptyset$. Two paths are *internally node-disjoint* if and only if they have no node in common at the possible exception of their terminal nodes (i.e. the two nodes that start and end the path). Formally, two paths $p_1 : u_1 \rightsquigarrow v_1$ and $p_2 : u_2 \rightsquigarrow v_2$ are internally disjoint if and only if $p_1 \cap p_2 \subseteq \{u_1, u_2, v_1, v_2\}$ holds. We recall that the container problem (a.k.a. the node-to-node disjoint paths routing problem) is about selecting internally disjoint paths between a pair of distinct nodes.

Definition 5 A path p connecting a node u to a node v satisfies the constraint $\gamma_i = (i, i+1)$ if and only if each node of p satisfies γ_i. We write $u \overset{\gamma_i}{\rightsquigarrow} v$, or simply $u \overset{\gamma}{\rightsquigarrow} v$.

So, as the Hamming distance between any two adjacent nodes in a hypercube is equal to one, a path cannot satisfy a 2-constraint other than that of the form $(i, i+1)$ (or $(i, i-1)$, which is equivalent). If the path $u \overset{\gamma}{\rightsquigarrow} v$ connecting the nodes u and v while satisfying a constraint γ_i has been generated by a shortest-path routing algorithm, we write $u \overset{\gamma, \text{spr}}{\rightsquigarrow} v$ to indicate that it is a shortest path.

We conclude this section by recalling the hypercube shortest-path routing algorithm with bit constraint of [13] in the following theorem.

Theorem 1 ([13]) *In a Q_n, given a bit constraint $\gamma_i = (i, i+1)$ and any two distinct nodes s and d that satisfy γ_i, we can select a shortest path $s \overset{\gamma}{\rightsquigarrow} d$ (i.e. of length $H(s, d)$) satisfying γ_i in $O(H(s, d))$ optimal time.*

Note that this algorithm is referred to as HC-SPR thereafter.

3 Node-to-node Disjoint Paths Routing Algorithm with γ_i Constraint

First, for a node $u \in Q_n$ satisfying $\gamma_i = (i, i+1)$, let us discuss the number of its neighbours that satisfy γ_i. If $w(u) = i + 1$, then u has $i + 1$ neighbours satisfying the constraint. If $w(u) = i$, then u has $n - i$ neighbours satisfying the constraint. Therefore, in a Q_n, given two nodes s, d satisfying γ_i, we can select at most $k \leq \min(n - i, i + 1)$ internally node-disjoint paths $s \overset{\gamma}{\rightsquigarrow} d$ that satisfy γ_i (this is an application of Menger's theorem [20]). In addition, one can note that in the case $i = 0$, the maximum number of disjoint paths that can be selected is $\min(n - i, i + 1) = 1$ and thus it is more efficient to apply the shortest-path routing algorithm of Theorem 1. So, let us assume that $i \geq 1$. Pseudo-code is given in Algorithm 1 with sub-cases in Algorithms 2 and 3.

3.1 Algorithm Description

If $n - i = 0$, the constraint $\gamma_i = (i, i + 1)$ cannot be satisfied as $i + 1 > n$. If $n - i = 1$, the constraint $\gamma_i = (i, i + 1)$ implies that only the nodes of weights n and $n - 1$ can be selected, thus $H(s, d) \leq 2$. The problem in this special case is solved as follows. If $H(s, d) = 1$, there exists only one path $s \overset{\gamma}{\rightsquigarrow} d$: the path of length one $s \overset{\gamma, \mathrm{spr}}{\rightsquigarrow} d = s \to d$. If $H(s, d) = 2$, there exists only one path $s \overset{\gamma}{\rightsquigarrow} d$: the path of length two $s \overset{\gamma, \mathrm{spr}}{\rightsquigarrow} d = s \to u \to d$ with u the unique node of Q_n of weight n. So, we can now assume that $n - i \geq 2$.

The main idea of this algorithm is to follow a divide-and-conquer approach by solving the problem recursively in one of the two subcubes Q_{n-1}^0 and Q_{n-1}^1 of the original network Q_n. The base case of this induction process is $k = 1$, with $k \leq \min(n - i, i + 1)$ the number of paths to find, decremented at each recursive call. This base case $k = 1$ induces either $i = 0$ or $i = n - 1$, and each of these two cases induces the selection of one single path (shortest) as already discussed. Let us distinguish two cases.

3.1.1 Case 1: $w(s) = i$

We proceed in several main steps as follows. Pseudo-code is given in Algorithm 2.

Step 1   Find a bit position δ ($0 \leq \delta \leq n - 1$) such that the δth bit of s is set to 0 and the δth bit of d is set to 1.
Reducing the hypercube Q_n along this bit position δ, we obtain the two subcubes Q_{n-1}^0 and Q_{n-1}^1, and $s \in Q_{n-1}^0, d \in Q_{n-1}^1$.

Step 2   Select the path of length one $s \in Q_{n-1}^0 \to s' \in Q_{n-1}^1$ with s' the unique neighbour of s in Q_{n-1}^1. Since $w(s) = i$ and the δth bit of s is 0, s' satisfies γ_i.

Step 3   Select the edge $s \to s''$ with s'' the unique neighbour of s in Q_{n-1}^1. Select the $k - 1$ neighbours $v_1, v_2, \ldots, v_{k-1}$ of s in Q_{n-1}^0 that satisfy γ_i. For an arbitrary bit position z such that the zth bit of s is set to 1, select the $k - 1$ nodes $u_1, u_2, \ldots, u_{k-1}$ in Q^0 with $u_j = v_j \oplus 2^z$. Then select the edges $u_j \to u'_j$ in Q_{n-1}^1. The nodes $s'', u'_1, u'_2, \ldots, u'_{k-1}$ are adjacent to the node $s' = (s \oplus 2^z) \oplus 2^\delta$.

Step 4   Apply this algorithm recursively in Q_{n-1}^1 to find $k - 1$ internally disjoint paths $s' \overset{\gamma}{\rightsquigarrow} d$ satisfying $\gamma_{i-1} = (i - 1, i)$.

Now, we distinguish two sub-cases depending on the value of $w(d)$.

Case 1.A: $w(d) = i + 1$.
The configuration in this case is given in Fig. 3.

Step 5   Select the edge $d \to d'$ with d' the unique neighbour of d in Q_{n-1}^0. Find a path $s \overset{\gamma}{\rightsquigarrow} d'$ in Q_{n-1}^0 as follows. Select a shortest path $s \overset{\gamma, \mathrm{spr}}{\rightsquigarrow} d'$ in Q_{n-1}^0.

Fig. 3 Illustration of the
case $w(s) = i$, $w(d) = i + 1$

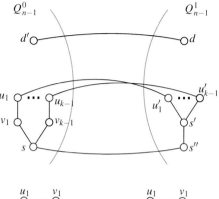

Fig. 4 Possible collisions
between the path $s \overset{\gamma}{\rightsquigarrow} d'$ and
a path $s \overset{\gamma}{\rightsquigarrow} u_j$ in Q^0_{n-1}: part
of $s \overset{\gamma}{\rightsquigarrow} u_j$ may be discarded

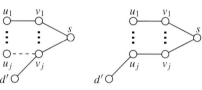

Let v be the closest node to d' on that path such that v is already included in a path $s \rightarrow v_j \rightarrow u_j$, say $s \rightarrow v_1 \rightarrow u_1$. So, s is connected to d' with the path $s \overset{\gamma}{\rightsquigarrow} v \overset{\gamma, \text{spr}}{\rightsquigarrow} d'$ with $s \overset{\gamma}{\rightsquigarrow} v$ a sub-path of $s \rightarrow v_1 \rightarrow u_1$. See Fig. 4.

Step 6   Assume without loss of generality that d' is connected to s in Q^0_{n-1} via $v_1 \in N(s)$; thus the edge $u_1 \rightarrow u'_1$ cannot be selected. The paths selected in Q^1_{n-1} are connected to s as follows.

First assume that the edge $s' \rightarrow s''$ is included in one of the selected paths. So, for this particular path, simply replace the edge $s' \rightarrow s''$ by $s \rightarrow s''$. Assume without loss of generality that there exists a node $u'_w \in Q^1_{n-1}$ neighbour of s' that is not included in any path selected in Q^1_{n-1}. If there is no such node, it means that a path $s' \overset{\gamma}{\rightsquigarrow} d$ selected in Q^1_{n-1} includes two nodes of $N(s')$, say u'_1, u'_2, and thus that path can be shortcut from $s' \overset{\gamma}{\rightsquigarrow} u'_1 \overset{\gamma}{\rightsquigarrow} u'_2 \rightarrow d$ to $s' \overset{\gamma}{\rightsquigarrow} u'_1 \rightarrow d$, freeing such a node u'_w (here $u'_w = u'_2$).

If the edge $s' \rightarrow u'_1$ is included in one of the selected paths, that path $s' \rightarrow u'_1 \overset{\gamma}{\rightsquigarrow} d$ is modified to $s \rightarrow v_w \rightarrow u_w \rightarrow u'_w \rightarrow s' \rightarrow u'_1 \overset{\gamma}{\rightsquigarrow} d$. Each of all other paths $s' \rightarrow u'_j \overset{\gamma}{\rightsquigarrow} d$ is modified to $s \rightarrow v_j \rightarrow u_j \rightarrow u'_j \overset{\gamma}{\rightsquigarrow} d$. And otherwise, each path $s' \rightarrow u'_j \overset{\gamma}{\rightsquigarrow} d$ is modified to $s \rightarrow v_j \rightarrow u_j \rightarrow u'_j \overset{\gamma}{\rightsquigarrow} d$, with the exception in the case $s = d'$ (i.e. $d = s''$) that the path $s' \rightarrow d$ of length 1 is modified to $s \rightarrow v_w \rightarrow u_w \rightarrow u'_w \rightarrow s' \rightarrow d$ instead.

Assume the edge $s' \rightarrow s''$ is not included in one of the selected paths in Q^1_{n-1}. The path $s' \rightarrow u'_1 \overset{\gamma}{\rightsquigarrow} d$ is modified to $s \rightarrow s'' \rightarrow s' \rightarrow u'_1 \overset{\gamma}{\rightsquigarrow} d$, and the other $k - 2$ paths $s' \rightarrow u'_j \overset{\gamma}{\rightsquigarrow} d$ ($2 \leq j \leq k - 1$) are modified to $s \rightarrow v_j \rightarrow u_j \rightarrow u'_j \overset{\gamma}{\rightsquigarrow} d$.

Fig. 5 Illustration of the case $w(s) = w(d) = i$

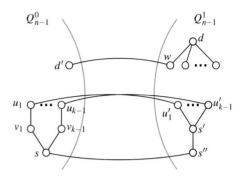

Case 1.B: w(d) = i.
  The configuration in this case is given in Fig. 5.

Step 5   Assume without loss of generality that there exists a node $w \in Q^1_{n-1}$ neighbour of d that is not included in any path selected in Q^1_{n-1}. If there is no such node, it means that a path $s' \xrightarrow{\gamma} d$ selected in Q^1_{n-1} includes two nodes of $N(d)$, say w_1, w_2, and thus that path can be shortcut from $s' \xrightarrow{\gamma} w_1 \xrightarrow{\gamma} w_2 \to d$ to $s' \xrightarrow{\gamma} w_1 \to d$, freeing such a node w (here $w = w_2$).
  Select the path $d \to w \to d'$ with d' the unique neighbour of w in Q^0_{n-1}. Find a path $s \xrightarrow{\gamma} d'$ in Q^0_{n-1} as in Step 5 of Case A.

Step 6   Similarly, the path $s \xrightarrow{\gamma} d'$ in Q^0_{n-1} may trigger rerouting in Q^1_{n-1} around s'. This is handled as in Step 6 of Case A.

3.1.2 Case 2: $w(s) = i + 1$

We first proceed in two steps before reducing this case to Case 1 (i.e. Sect. 3.1.1). Pseudo-code is given in Algorithm 3.

Step 1   Find a bit position δ ($0 \le \delta \le n - 1$) such that the δth bit of s is set to 1 and the δth bit of d is set to 0.
  Reducing the hypercube Q_n along this bit position δ, we obtain the two subcubes Q^0_{n-1} and Q^1_{n-1}, and $s \in Q^1_{n-1}, d \in Q^0_{n-1}$.

Step 2   Select the path of length one $s \in Q^1_{n-1} \to s' \in Q^0_{n-1}$ with s' the unique neighbour of s in Q^0_{n-1}. Since $w(s) = i + 1$ and the δth bit of s is set to 1, s' satisfies γ_i.

Then, the case $w(s) = i + 1, w(d) = i$ is solved similarly to the case $w(s) = i$, $w(d) = i+1$ (Case 1.A) by exchanging the roles of s and d, and the case $w(s) = i+1$, $w(d) = i + 1$ is solved similarly to the case $w(s) = i, w(d) = i$ (Case 1.B) with the differences that

- in Step 3, the bit position z is selected such that the zth bit of s is set to 0;
- in Step 4, the algorithm is applied recursively in Q_{n-1}^0, the constraint considered thus remaining $\gamma_i = (i, i+1)$;
- in Step 5, a shortest path routing algorithm is applied in Q_{n-1}^1, the constraint considered thus becoming $\gamma_{i-1} = (i-1, i)$.

Algorithm 1 HC-CONTAINER(Q_n, i, k, s, d)

Input: A Q_n, a bit constraint $\gamma_i = (i, i+1)$, k the number of paths to find ($k \le \min(n-i, i+1)$), a source node s and a destination node d.

Output: k internally node-disjoint paths $s \overset{\gamma}{\rightsquigarrow} d$ in Q_n satisfying γ_i.

1: **if** $k = 1$ **then**
2:    **return** HC-SPR(Q_n, i, s, d)
3: **else if** $w(s) = i$ **then**
4:    **return** CASE1(Q_n, i, k, s, d)
5: **else** // $w(s) = i+1$
6:    **return** CASE2(Q_n, i, k, s, d)
7: **end if**

3.2 Correctness and Complexities

In this section, we show the correctness of the algorithm of Sect. 3.1 and establish its worst case time and path length complexities.

Lemma 1 *The algorithm of Sect. 3.1 is correct and always terminates.*

Proof We recall that a path of length one $s \rightarrow s'$ is selected, with s, s' in distinct subcubes. We start by showing the existence of a reduction bit δ.

Assume $w(s) = i$. Suppose there is no bit position δ with the δth bit of s set to 0 and the δth bit of d set to 1. If $w(d) = i$, this supposition implies that $s = d$, which is a contradiction. If $w(d) = i+1$, this supposition implies that all the bits of d corresponding to the positions of the $n-i$ bits of s set to 0 are also set to 0, and thus that d has $n-i$ bits set to 0 and $i+1$ bits set to 1. This is a contradiction since d of n bits ($d \in Q_n$).

Assume $w(s) = i+1$. Suppose there is no bit position δ with the δth bit of s set to 1 and the δth bit of d set to 0. If $w(d) = i+1$, this supposition implies that $s = d$, which is a contradiction. If $w(d) = i$, this supposition implies that all the bits of s corresponding to the positions of the $n-i$ bits of d set to 0 are also set to 0, and thus that s has $n-i$ bits set to 0 and $i+1$ bits set to 1. This is a contradiction since s of n bits ($s \in Q_n$).

Algorithm 2 CASE1(Q_n, i, k, s, d)

Input: A Q_n, a bit constraint $\gamma_i = (i, i+1)$, k the number of paths to find ($k \leq \min(n-i, i+1)$), a
     source node s with $w(s) = i$ and a destination node d.

Output: $k = \min(n-i, i+1)$ internally node-disjoint paths $s \overset{\gamma}{\leadsto} d$ in Q_n satisfying γ_i.

1:   $R := \emptyset$; // Result set
2:   $\delta := \lfloor \log((s \oplus d) \& d) \rfloor$;
3:   $z := \lfloor \log s \rfloor$;
4:   $s'' := s \oplus 2^\delta$;
5:   $s' := s \oplus z \oplus 2^\delta$;
6:   // The simple cases $s' = d$ and $s'' = d$ are omitted.
7:   $\{v_1, v_2, \ldots, v_{k-1}\} := N(s) \cap Q_{n-1}^0$ sat. γ_i;
8:   $\{u_1, u_2, \ldots, u_{k-1}\} := \{v_1 \oplus z, v_2 \oplus z, \ldots, v_{k-1} \oplus z\}$;
9:   $\{u_1', u_2', \ldots, u_{k-1}'\} := \{u_1 \oplus 2^\delta, u_2 \oplus 2^\delta, \ldots, u_{k-1} \oplus 2^\delta\}$;
10: $P := \text{HC-CONTAINER}(Q_{n-1}^1, i-1, k-1, s', d)$;
11: **if** $w(d) = i+1$ **then**
12:     $d' := d \oplus 2^\delta$;
13:     $(p^* : s' \to u_x' \overset{\gamma}{\leadsto} d') := \text{HC-SPR}(Q_{n-1}^0, i, s, d')$;
14:     **for all** $p : s' \to u_j' \overset{\gamma}{\leadsto} d$ in P **do**
15:       **if** $u_j' = u_x'$ **then**
16:         **if** $\exists q \in P$ with $(s' \to s'') \in q$ **then**
17:           $u_w' := \{u_1', u_2', \ldots, u_j'\} \setminus \bigcup_{q \in P} q$;
18:           $R := R \cup \{s \to v_w \to u_w \to u_w' \to p\}$
19:         **else**
20:           $R := R \cup \{s \to s'' \to p\}$
21:         **end if**
22:       **else**
23:         **if** $u_j' = s''$ **then**
24:           $R := R \cup \{s \to (u_j' \overset{\gamma}{\leadsto} d) \subset p\}$
25:         **else**
26:           $R := R \cup \{s \to v_j \to u_j \to (u_j' \overset{\gamma}{\leadsto} d) \subset p\}$
27:         **end if**
28:       **end if**
29:     **end for**
30: **else** // $w(d) = i$
31:     $w := (N(d) \cap Q_{n-1}^1$ sat. $\gamma_i) \setminus \bigcup_{q \in P} q$;
32:     $d' := w \oplus 2^\delta$;
33:     $(p^* : s' \to u_x' \overset{\gamma}{\leadsto} d') := \text{HC-SPR}(Q_{n-1}^0, i, s, d')$;
34:     **for all** $p : s' \to u_j' \overset{\gamma}{\leadsto} d$ in P **do**
35:       **if** $u_j' = u_x'$ **then**
36:         **if** $\exists q \in P$ with $(s' \to s'') \in q$ **then**
37:           $u_w' := \{u_1', u_2', \ldots, u_j'\} \setminus \bigcup_{q \in P} q$;
38:           $R := R \cup \{s \to v_w \to u_w \to u_w' \to p\}$
39:         **else**
40:           $R := R \cup \{s \to s'' \to p\}$
41:         **end if**
42:       **else**
43:         **if** $u_j' = s''$ **then**

44:               $R := R \cup \{s \rightarrow (u'_j \overset{\gamma}{\rightsquigarrow} d) \subset p\}$

45:        **else**

46:               $R := R \cup \{s \rightarrow v_j \rightarrow u_j \rightarrow (u'_j \overset{\gamma}{\rightsquigarrow} d) \subset p\}$

47:        **end if**

48:     **end if**

49:   **end for**

50: **end if**

51: **return** $R \cup \{p^* \rightarrow d\}$

Algorithm 3 CASE2(Q_n, i, k, s, d)

Input: A Q_n, a bit constraint $\gamma_i = (i, i+1)$, k the number of paths to find ($k \leq \min(n-i, i+1)$), a source node s with $w(s) = i+1$ and a destination node d.

Output: $k = \min(n-i, i+1)$ internally node-disjoint paths $s \overset{\gamma}{\rightsquigarrow} d$ in Q_n satisfying γ_i.

1: **if** $w(d) = i$ **then**

2:    **return** HC-CONTAINER(Q_n, i, k, d, s)

3: **else**

4:    $R := \emptyset$; // *Result set*

5:    $\delta := \lfloor \log((s \oplus d) \& s) \rfloor$;

6:    $z := \lfloor \log \neg s \rfloor$;

7:    $s'' := s \oplus 2^\delta$;

8:    $s' := s \oplus z \oplus 2^\delta$;

9:    $\{v_1, v_2, \ldots, v_{k-1}\} := N(s) \cap Q^1_{n-1}$ sat. γ_i;

10:   $\{u_1, u_2, \ldots, u_{k-1}\} := \{v_1 \oplus z, v_2 \oplus z, \ldots, v_{k-1} \oplus z\}$;

11:   $\{u'_1, u'_2, \ldots, u'_{k-1}\} := \{u_1 \oplus 2^\delta, u_2 \oplus 2^\delta, \ldots, u_{k-1} \oplus 2^\delta\}$;

12:   $P := $ HC-CONTAINER(Q^0_{n-1}, i, $k-1$, s', d);

13:   $w := (N(d) \cap Q^1_{n-1}$ sat. $\gamma_i) \setminus \bigcup_{q \in P} q$;

14:   // *The simple cases $s' = d$ and $s'' = w$ are omitted.*

15:   $d' := w \oplus 2^\delta$;

16:   $(p^* : s' \rightarrow u'_x \overset{\gamma}{\rightsquigarrow} d') := $ HC-SPR(Q^1_{n-1}, $i-1$, s, d');

17:   **for all** $p : s' \rightarrow u'_j \overset{\gamma}{\rightsquigarrow} d$ in P **do**

18:     **if** $u'_j = u'_x$ **then**

19:       **if** $\exists q \in P$ with $(s' \rightarrow s'') \in q$ **then**

20:         $u'_w := \{u'_1, u'_2, \ldots, u'_j\} \setminus \bigcup_{q \in P} q$;

21:         $R := R \cup \{s \rightarrow v_w \rightarrow u_w \rightarrow u'_w \rightarrow p\}$

22:       **else**

23:         $R := R \cup \{s \rightarrow s'' \rightarrow p\}$

24:       **end if**

25:     **else**

26:       **if** $u'_j = s''$ **then**

27:         $R := R \cup \{s \rightarrow (u'_j \overset{\gamma}{\rightsquigarrow} d) \subset p\}$

28:       **else**

29:         $R := R \cup \{s \rightarrow v_j \rightarrow u_j \rightarrow (u'_j \overset{\gamma}{\rightsquigarrow} d) \subset p\}$

30:       **end if**

31:     **end if**

32:   **end for**

33:   **return** $R \cup \{p^* \rightarrow w \rightarrow d\}$

34: **end if**

Regarding the existence of available neighbours, rerouting feasibility, etc., a proof has already been given in the corresponding steps of Sect. 3 for more clarity.

Lemma 2 *The algorithm of Sect. 3.1 generates internally node-disjoint paths of lengths at most $n + 3k$ in $O(kn)$ time.*

Proof The paths generated by the algorithm of Sect. 3.1 are internally node-disjoint as shown by the algorithm description.

We now consider the maximum length of a generated path. A total of $k = \min(n - i, i + 1)$ paths are generated. First, one should note that exactly one hypercube reduction is required for each path to be generated. So, in total, the original hypercube Q_n is reduced k times. In other words, routing is performed inside hypercubes of successive dimensions $n, n - 1, \ldots, n - k$. When the base case condition $k = 1$ is satisfied a shortest-path routing algorithm is applied. This base case $k = 1$ actually means that either $i = 0$ and thus a path of length at most two is generated, or $i = n - 1$ and thus a path of length at most $n - k$ is generated. In addition, each hypercube reduction induces $3 - 1 = 2$ additional edges to connect the paths selected by induction inside one subcube to the node s located inside the other subcube. Rerouting triggered in Step 6 may induce an extra two edges to connect a path, say $u_1' \overset{\gamma}{\rightsquigarrow} d$, in the sub-cube of d to s via the special node $u_w' \in N(s')$ (precisely, the two extra edges are $u_w' \rightarrow s' \rightarrow u_1'$). Thus, at most four edges in total to be added at each reduction to connect paths in the sub-cube of d to s. Therefore, generated paths have lengths of at most $4k + (n - k) = n + 3k$ edges. The single path connecting d to s by application of a shortest-path routing algorithm inside the subcube of s requires at most two edges for the sub-path $d \overset{\gamma}{\rightsquigarrow} d'$ and at most $n - 1$ edges for the shortest path $d' \overset{\gamma, \text{spr}}{\rightsquigarrow} s$, thus requiring in total at most $n + 1$ edges.

Regarding the time complexity of the algorithm of Sect. 3.1, Steps 1 and 2 are both constant time $O(1)$. Step 3 is linear time $O(n)$. Let $T(n)$ be the time required to solve the problem in a Q_n. Step 4 is thus $T(n - 1)$ time. Steps 5 and 6 are both linear time $O(n)$. From this discussion, we obtain the equation $T(n) = T(n - 1) + O(n)$. Since we have exactly k hypercube reductions, the total time complexity of the proposed algorithm is $O(kn)$. □

So, we can summarise this discussion in the following theorem.

Theorem 2 *In a Q_n, given a bit constraint $\gamma_i = (i, i + 1)$ and any two distinct nodes s and d satisfying γ_i, we can select $k = \min(n - i, i + 1)$ internally node-disjoint paths $s \overset{\gamma}{\rightsquigarrow} d$ satisfying γ_i and of lengths at most $n + 3k$ in $O(kn)$ time.*

Proof This can be directly deduced from Lemmas 1 and 2. □

3.3 Routing Example

In a Q_5, given a bit constraint $\gamma_2 = (2, 3)$, a source node $s : 11010$ and a destination node $d : 10101$, an execution trace of the algorithm of Sect. 3 is given in Table 1.

Table 1 Node-to-node disjoint paths routing example in a Q_5 with $\gamma_2 = (2, 3)$ bit constraint

n	2^δ	s	d	s'	d'	$\in Q^0_{n-1}$	$\in Q^1_{n-1}$
5	8	11010	10101	10110	-	d	s
Selection of sub-path $s = 11010 \rightarrow 1010 \rightarrow 1110$.							
Induction on Q^0_4.							
4	2	10110	10101	10101	-	d, s'	s
Selection of sub-path $s = 10110 \rightarrow 10010 \rightarrow 10011$.							
Induction on Q^0_3: $s' = d$ and thus selection of							
$s = 10110 \rightarrow 10010 \rightarrow 10011 \rightarrow 10001 \rightarrow 10101 = d$.							
4	2	10110	10101	10101	00111	d, s'	s, d'
Selection of $s = 10110 \rightarrow 00110 \rightarrow 00111 \rightarrow 00101 \rightarrow 10101 = d$							
(by SPR routing in Q^1_3).							
5	8	11010	10101	10110	11100	d, s'	s, d'
Selection of $s = 11010 \rightarrow 11000 \rightarrow 11100 \rightarrow 10100 \rightarrow 10101 = d$							
(by SPR routing in Q^1_4),							
$s = 11010 \rightarrow 01010 \rightarrow 01110 \rightarrow 00110 \rightarrow 00111 \rightarrow 00101 \rightarrow 10101 = d$,							
and $s = 11010 \rightarrow 10010 \rightarrow 10011 \rightarrow 10001 \rightarrow 10101 = d$							
(by joining sub-paths).							

As a result, the following three internally node-disjoint paths, all satisfying γ_2, are selected:

- $s = 11010 \rightarrow 11000 \rightarrow 11100 \rightarrow 10100 \rightarrow 10101 = d$
- $s = 11010 \rightarrow 10010 \rightarrow 10011 \rightarrow 10001 \rightarrow 10101 = d$
- $s = 11010 \rightarrow 01010 \rightarrow 01110 \rightarrow 00110 \rightarrow 00111 \rightarrow 00101 \rightarrow 10101 = d$

4 Empirical Evaluation

We conducted an empirical evaluation of the proposed hypercube node-to-node disjoint paths routing algorithm to inspect its practical behaviour, that is how the algorithm performs in average. Our objective is also to compare these experimental results to the theoretical estimations of Sect. 3.2.

We have implemented the algorithm of Sect. 3 using the the Scheme functional programming language [21]. Then, in a hypercube of dimension n ($4 \leq n \leq 8$) and a bit constraint $\gamma_2 = (2, 3)$, we have used this implementation to solve 1,000 random instances of the container problem for each value of n. Values of n smaller than 4

Fig. 6 Maximum path
length and average maximum
path length with standard
deviation for each value of n
($i = 2$ and $4 \leq n \leq 8$)

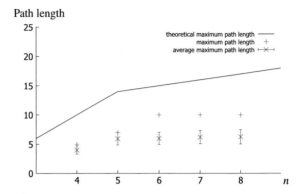

are ignored since at most $k = \min(n - i, i + 1)$ disjoint paths can be found, that is at most one path for $n < 4$, and thus a shortest-path routing algorithm would suffice.

In this experiment, we have measured the maximum path length obtained for each value of n, and also for each value of n the average of the 1,000 maximum path lengths, each average value being obtained when solving one instance of the container problem. The results are illustrated in Fig. 6. To facilitate comparison with the theoretical estimation, we have additionally plotted the theoretical maximum path length as estimated in Lemma 2.

One can see that our theoretical estimation regarding the maximum path length is slightly pessimistic as experimentation results show a small gap between the theoretical maximum path length and the empirical one. We may thus be able to refine our theoretical estimations.

5 Conclusions

Due to their simplicity for both hardware and software implementation, hypercubes are very popular as interconnection network of massively parallel systems. Disjoint paths routing is one mainstream method to avoid notorious parallel processing problems such as deadlocks. We have proposed in this paper an algorithm solving the container problem with bit constraint inside a hypercube Q_n. Given a bit constraint $\gamma_i = (i, i + 1)$ and any two distinct nodes s, d satisfying γ_i, the algorithm selects $k = \min(n - i, i + 1)$ internally node-disjoint paths $s \xrightarrow{\gamma} d$ satisfying γ_i. The paths selected are of lengths at most $n + 3k$, and the time complexity of this routing algorithm is $O(kn)$. Therefore, by enforcing a bit constraint when routing, this algorithm provides the ability to obtain several sets of disjoint paths between several node pairs, each pair satisfying a distinct bit constraint. Conventional routing algorithms, even disjoint paths routing algorithms, are not able to provide such result.

Future works first include experimenting further with parameters of higher values. Then, we are planning to extend this research to solve the node-to-set disjoint paths routing problem with bit constraint in a hypercube. Also, enhanced fault tolerance, such as cluster-fault tolerance, is an interesting future development.

Acknowledgments The authors sincerely thank the reviewers for their insightful comments. This study was partly supported by a Grant-in-Aid for Scientific Research (C) of the Japan Society for the Promotion of Science under Grant No. 25330079.

References

1. Saad, Y., Schultz, M.H.: Topological properties of hypercubes. IEEE Trans. Comput. **37**(7), 867–872 (1988)
2. Seitz, C.L.: The cosmic cube. Commun. ACM **28**(1), 22–33 (1985)
3. TOP500. List. http://top500.org/list/2014/06/, June 2014. Last accessed July 2014
4. Li, Y., Peng, S., Chu, W.: Efficient collective communications in dual-cube. J. Supercomput. **28**(1), 71–90 (2004)
5. Li, Y., Peng, S., Chu, W.: Metacube - a versatile family of interconnection networks for extremely large-scale supercomputers. J. Supercomput. **53**(2), 329–351 (2010)
6. Malluhi, Q.M., Bayoumi, M.A.: The hierarchical hypercube: a new interconnection topology for massively parallel systems. IEEE Trans. Parallel Distrib. Syst. **5**(1), 17–30 (1994)
7. Ghose, K., Desai, K.R.:The HCN: a versatile interconnection network based on cubes. In: Proceedings of the 1989 ACM/IEEE Conference on Supercomputing, pp. 426–435. Reno, NV, USA, November 12–17 (1989)
8. Gao, S., Novick, B., Qiu, K.: From Hall's matching theorem to optimal routing on hypercubes. J. Comb. Theory Ser. B **74**, 291–301 (1998)
9. Sinanoglu, O., Karaata, M.H., AlBdaiwi, B.: An inherently stabilizing algorithm for node-to-node routing over all shortest node-disjoint paths in hypercube networks. IEEE Trans. Comput. **59**(7), 995–999 (2010)
10. Bossard, A., Kaneko, K.: Time optimal node-to-set disjoint paths routing in hypercubes. J. Inf. Sci. Eng. **30**(4), 1087–1093 (2014)
11. Gu Q.-P., Okawa S., Peng S.: Set-to-set fault tolerant routing in hypercubes. IEICE Trans. Fundam. **E79-A**(4):483–488 (1996)
12. Gu, Q.-P., Peng, S.: An efficient algorithm for the k-pairwise disjoint paths problem in hypercubes. J. Parallel Distrib. Comput. **60**(6), 764–774 (2000)
13. Bossard, A., Kaneko, K.: On hypercube routing and fault tolerance with bit constraint. In: Proceedings of the Second International Symposium on Computing and Networking, pp. 40–49. Shizuoka City, Japan, December 10–12 (2014)
14. Li, Y., Peng, S., Chu, W.: Disjoint paths in metacube. In: Proceedings of the IASTED International Conference on Parallel and Distributed Computing and Systems, pp. 43–50. Marina del Rey, CA, USA, November 3–5 (2003)
15. Murugesan, S.: Harnessing Green IT: principles and practices. IT Prof. **10**(1), 24–33 (2008)
16. Chen, J., Kanj, I.A., Wang, G.: Hypercube network fault tolerance: a probabilistic approach. J. Interconnect. Netw. **6**(1), 17–34 (2005)
17. Dietzfelbinger, M., Madhavapeddy, S., Sudborough, I.H.: Three disjoint path paradigms in star networks. In: Proceedings of the Third IEEE Symposium on Parallel and Distributed Processing, pp. 400–406. Dallas, TX, USA, December 2–5 (1991)
18. Suzuki, Y., Kaneko, K.: An algorithm for node-disjoint paths in pancake graphs. IEICE Trans. Inf. Syst. **E86-D**(3):610–615 (2003)

19. Kaneko, K., Sawada, N.: An algorithm for node-to-node disjoint paths problem in burnt pancake graphs. IEICE Trans. Inf. Syst. **E90-D**(1):306–313 (2007)
20. Menger, K.: Zur allgemeinen Kurventheorie. Fundamenta Mathematicae **10**, 96–115 (1927)
21. Findler, R.B., Clements, J., Flanagan, C., Flatt, M., Krishnamurthi, S., Steckler, P., Felleisen, M.: DrScheme: a programming environment for scheme. J. Funct. Prog. **12**(2), 159–182 (2002)

Algorithms for Removing Node Overlaps with Some Basis Nodes

Noboru Abe, Hiroaki Oh and Kouhei Inoue

Abstract Graphs are used to represent various types of structures. When the nodes of a graph are drawn by non-zero sized graphical features, it is important to avoid node overlaps. We propose three heuristic algorithms to remove node overlaps in graphs with several tens of nodes by refining a previously proposed algorithm, i.e., the force-transfer algorithm.

Keywords Graph layout · Overlapping nodes · Force transfer · Layout adjustment

1 Introduction

This paper considers drawing graphs with nodes that are non-zero sized graphical features, such as axis-parallel rectangles. Note that avoiding node overlaps with minimum layout area is an important problem for such a graph. First, one can obtain a graph layout using algorithms that do not consider the node size [1–5]. Next, the nodes of that graph can be replaced with non-zero sized nodes. Then, node overlaps can be removed. In this case, the moved distances of nodes must be minimized to preserve the layout esthetics. Furthermore, in dynamically changing graphs, it is important to preserve the mental map [6] of the original graph.

Several algorithms have been proposed for this problem, such as the force-scan algorithm (FSA) [6] and its variants [7, 8]. The FSA moves each node by scanning the layout horizontally and vertically. In addition, algorithms that use a Voronoi diagram or Delaunay triangulation [9, 10], algorithms that employ a quadratic programming [11, 12], and algorithms that expand the layout until overlaps are removed [6, 11] were proposed.

N. Abe (✉) · H. Oh · K. Inoue
Faculty of Information and Communication Engineering,
Osaka Electro-Communication University, 18-8 Hatsu-cho, Neyagawa-shi,
Osaka-fu 572–8530, Japan
e-mail: abe@isc.osakac.ac.jp

© Springer International Publishing Switzerland 2016   93
R. Lee (ed.), *Software Engineering, Artificial Intelligence, Networking
and Parallel/Distributed Computing 2015*, Studies in Computational Intelligence 612,
DOI 10.1007/978-3-319-23509-7_7

In this study, we propose three heuristic algorithms to remove node overlaps in a graph with several tens of nodes by refining the force-transfer algorithm (FTA) [8]. Note that this study does not consider the edges of a graph. To handle edges, we can simply create a span between nodes. In such cases, edge-node intersections may occur. However, this problem can be readily solved by bending edges [13]. An algorithm that avoids edge-node intersections without bending edges has been proposed [14]. However, this algorithm does not consider preserving the mental map.

The remainder of this paper is organized as follows. In Sect. 2, we define the problem of removing node overlaps and briefly explain the FTA. In Sect. 3, we propose three algorithms for the problem. We show experimental results in Sect. 4 and conclude the paper in Sect. 5.

2 Problem Definition and Force-Transfer Algorithm

First, we define the removing node overlaps problem. Then, we briefly explain the FTA [8].

2.1 Problem Definition

We assume that a graph G has a set of nodes denoted by $V = \{1, 2, \ldots, |V|\}$ and some of the nodes overlap. Let (x_i^0, y_i^0) be the center of node i, and let (x_i^1, y_i^1) and (x_i^2, y_i^2) be the lower left and upper right corner coordinates, respectively (Fig. 1).

Huang et al. [8] refer to the *neighbor nodes* of node q, denoted by NN(q), as the set of nodes that overlap q directly. For the graph shown in Fig. 2, we have NN(q) = $\{h, j, k\}$. In addition, they define the *left, right, up*, and *down neighbor nodes* of q as follows:

Fig. 1 An example of a node

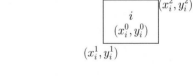

Fig. 2 An example of neighbor nodes

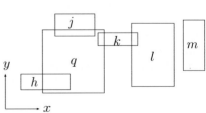

Fig. 3 An example of
a conflict graph

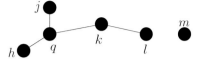

$$\text{LNN}(q) = \{i|(x_i^1 < x_q^1 \lor x_i^2 < x_q^2) \land i \in \text{NN}(q)\},$$
$$\text{RNN}(q) = \{i|(x_i^1 \geq x_q^1 \lor x_i^2 \geq x_q^2) \land i \in \text{NN}(q)\},$$
$$\text{UNN}(q) = \{i|(y_i^1 \geq y_q^1 \lor y_i^2 \geq y_q^2) \land i \in \text{NN}(q)\},$$
$$\text{and DNN}(q) = \{i|(y_i^1 < y_q^1 \lor y_i^2 < y_q^2) \land i \in \text{NN}(q)\}$$

For the graph shown in Fig. 2, $\text{LNN}(q) = \{h, j\}$, $\text{RNN}(q) = \{j, k\}$, $\text{UNN}(q) = \{h, j, k\}$, and $\text{DNN}(q) = \{h, k\}$. Note that we add conditional expressions for x_i^2 and y_i^2 to these definitions because the original definitions cannot frequently determine the most efficient move direction. For example, in Fig. 2, it is efficient to move h to the bottom of q to remove the overlap of h and q. However, the original definitions miss this direction because the original $\text{DNN}(q)$ does not contain h.

Here, we introduce the *conflict graph* from graph G. Figure 3 shows the *conflict graph* for the graph shown in Fig. 2. The *conflict graph* has nodes with one-to-one correspondence with the nodes of G and edges if and only if the corresponding nodes of G overlap each other.

The set of all nodes of the connected component containing q of a *conflict graph* is denoted by $CC(q)$. Huang et al. [8] refer to the set of all nodes in $CC(q) - \{q\}$ as the *transfer neighbor nodes* of node q, denoted by $\text{TNN}(q)$. For the graph in Fig. 2, $\text{TNN}(q) = \{h, j, k, l\}$. In addition, Huang et al. define the *left, right, up,* and *down transfer neighbor nodes* of q as follows:

$$\text{TLNN}(q) = \{i|(x_i^1 < x_q^1 \lor x_i^2 < x_q^2) \land i \in \text{TNN}(q)\},$$
$$\text{TRNN}(q) = \{i|(x_i^1 \geq x_q^1 \lor x_i^2 \geq x_q^2) \land i \in \text{TNN}(q)\},$$
$$\text{TUNN}(q) = \{i|(y_i^1 \geq y_q^1 \lor y_i^2 \geq y_q^2) \land i \in TNN(q)\},$$
$$\text{and TDNN}(q) = \{i|(y_i^1 < y_q^1 \lor y_i^2 < y_q^2) \land i \in \text{TNN}(q)\}$$

Note that for the graph in Fig. 2, $\text{TLNN}(q) = \{h, j\}$, $\text{TRNN}(q) = \{j, k, l\}$, $\text{TUNN}(q) = \{h, j, k, l\}$, and $\text{TDNN}(q) = \{h, k\}$.

The objective of the defined problem is to remove node overlaps by adjusting node positions. Huang et al. [8] present four measures, $\lambda_1, \lambda_2, \lambda_3$, and f_{cost}, to evaluate the quality of an adjustment.

λ_1, defined as $\lambda_1 = n/|V|$ where n is the number of adjusted nodes, counts the number of repositioned nodes.

λ_2, defined as $\tau/(|V|(|V|-1))$, measures the change of the relative positions of nodes. Here, τ is the number of node pairs whose relative positions have changed in the x-axis and y-axis directions in an adjustment.

λ_3 measures the change of the layout area and is defined as $\lambda_3 = 1 - W/W'$, where W denotes the minimal area of the bounding rectangle of the original graph layout and W' denotes that of the adjusted layout.

f_{cost}, defined as the total sum of the moved distances of each node in the x- and y-directions, measures the adjusted distances of all nodes.

The removing node overlaps problem is considered a multi-objective optimization problem to minimize the above four measures under the condition that all nodes must not overlap.

2.2 Force-Transfer Algorithm

Huang et al. [8] proposed the FTA to remove node overlaps. The FTA is proposed by refining the FSA [6], and its above-mentioned adjustment quality measures are significantly improved. The FTA begins a scan from a basis node called a *seed node*. Note that the *seed node* affects the adjusted layout significantly [8]. In general, Huang et al. selected the leftmost node as the *seed node* in their experimentation.

Starting from the seed node, the FTA executes four scans and moves overlapping nodes located to the right, left, above, and below of the seed node. The FTA is described as follows:

[FTA]

(1) Sort all nodes according to their x_i^1 coordinates.
(2) Select a node or use the leftmost node as the seed node s.
(3) Execute the Right Horizontal Transfer procedure.
(4) Execute the Left Horizontal Transfer procedure.
(5) Execute the Up Vertical Transfer procedure.
(6) Execute the Down Vertical Transfer procedure.

The Right Horizontal Transfer procedure is described as follows:

Right Horizontal Transfer:

(a) For each node $i = s, s + 1, s + 2, \ldots, |V|$, execute the following.

(a-1) Find RNN(i) and TRNN(i).
(a-2) For each node $j \in$ RNN(i), execute the following.
(a-2-1) Calculate $f_{ij}^x = x_i^2 - x_j^1$ and $f_{ij}^y = min\{|y_i^2 - y_j^1|, |y_i^1 - y_j^2|\}$.
(a-2-2) If $f_{ij}^x \le f_{ij}^y$, then $x_k^0 = x_k^0 + f_{ij}^x$ for each node $k \in$ TRNN(i).
(a-2-3) Update RNN(i) and TRNN(i).

We perform other three Transfer procedures similarly, after re-sorting all nodes.

3 Proposed Algorithms

The FTA was primarily proposed to handle graphs with few nodes. For graphs with many nodes, Huang et al. [8] provided a new version of the algorithm with virtual nodes. Each virtual node contains all nodes in a connected component of the *conflict graph*. In Fig. 4a, the dashed rectangles represent virtual nodes. On the first abstraction level, the FTA is applied to remove node overlaps within each virtual node (Fig. 4b), and then, on the next abstraction level, the FTA is applied to remove overlaps of virtual nodes. This procedure iteratively executes until all overlaps have been removed.

When the *conflict graph* has a big connected component, the above-mentioned virtual node version of the FTA does not work well. For this reason, Huang et al. provided a more generalized version of virtual node version of the FTA that uses grid cells for abstraction levels. However, the virtual node version may yield unnecessary node movement. For example, as can be seen in Fig. 4b, the two lower virtual nodes are moved to remove overlap even though the nodes in the virtual nodes do not overlap. The grid cell version has the same problem. In addition, determining grid cell size is problematic in the grid cell version.

We want to propose algorithms that are independent of grid cell size. In the case that the normal version of the FTA is applied for graphs with many nodes, we found that some overlaps remain. These overlaps occasionally remain for graphs with few nodes. Here, we briefly describe two typical cases. Figure 5 shows the first case. Node a has moved up to remove the overlap of nodes a and b in step (5) of the FTA; however, nodes a and c now overlap. To remove this overlap, a Right or Left Horizontal Transfer procedure is required. However, these procedures can no longer be executed because steps (3) and (4) for these procedures are already finished.

If the selected seed node s is not the leftmost, rightmost, topmost, and bottommost node, the second case occurs (Fig. 6). To remove the overlap of a and b, a Right or Left

(a)   **(b)**

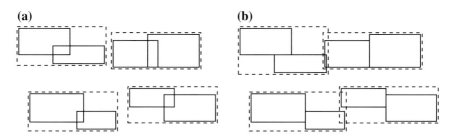

Fig. 4 An example of virtual nodes

Fig. 5 An example of
remaining overlap (case 1)

Fig. 6 An example of
remaining overlap (case 2)

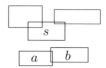

Horizontal Transfer procedure is required. However, the Right Horizontal Transfer procedure omits a because it does not exist on the right side of s, and the Left Horizontal Transfer procedure omits b because it does not exist on the left side of s.

We refine the FTA slightly to avoid such overlaps. We refer to this algorithm as Method 1. The Method 1 algorithm is described as follows:

[Method 1]

(1) Iterate the following steps until all node overlaps have been removed.

    (1-a) Sort all nodes according to their x_i^1 coordinates.
    (1-b) Select the leftmost node as the seed node s.
    (1-c) Execute the Right Horizontal Transfer procedure.
    (1-d) Sort all nodes according to their x_i^2 coordinates.
    (1-e) Select the rightmost node as the seed node s.
    (1-f) Execute the Left Horizontal Transfer procedure.
    (1-g) Execute Vertical Transfer procedures in the same manner.

We also propose two additional algorithms that are tailored for particular purposes by refining Method 1. Method 2 attempts to minimize the layout area. The Method 2 algorithm is described as follows:

[Method 2]

(1) Sort all nodes according to their x_i^1 coordinates.
(2) Select the leftmost node as the seed node s.
(3) Let $i = s$. Execute only steps (a-1) and (a-2) of the Right Horizontal Transfer procedure.
(4) Sort all nodes according to their x_i^2 coordinates.
(5) Select the rightmost node as the seed node s.
(6) Let $i = s$. Execute only steps (a-1) and (a-2) of the Left Horizontal Transfer procedure.
(7) Execute Vertical Transfer procedures in the same manner.
(8) Execute all steps of Method 1.

Initially, Method 2 attempts to move nodes inside the layout to remove overlaps of the leftmost, rightmost, topmost, and bottommost nodes. Note that steps (3), (6), and (7) of Method 2 move only the *transfer neighbor nodes* of s to avoid expanding the layout unnecessarily.

The third algorithm attempts to minimize the adjusted distances of nodes and preserve the mental map. Note that nodes that are distant from s tend to move a great distance. Thus, this algorithm, which we refer to as Method 3, attempts to avoid this by selecting the seed node appropriately. The Method 3 algorithm is described as follows:

[Method 3]

(1) Sort all nodes according to their x_i^1 coordinates.
(2) Let mid be the $\lfloor |V|/2 \rfloor$th node. Select seed node s by executing the following steps.

    (2-a) Find the connected component C of the conflict graph that contains the node mid.

    (2-b) For each node $i \in C$ of the conflict graph, calculate the shortest path sp_{ij} to node $j \in C$.

    (2-c) For each node $i \in C$ of the conflict graph, find the maximum value max_sp_i of sp_{ij}.

    (2-d) Select node $i \in C$ with the minimum max_sp_i value as seed node s.

(3) Execute the Right Horizontal Transfer procedure.
(4) Sort all nodes according to their x_i^2 coordinates.
(5) Execute the Left Horizontal Transfer procedure without reselecting seed node s.
(6) Execute Vertical Transfer procedures in the same manner.
(7) Execute all steps of Method 1.

For the graphs shown in Figs. 2 and 3, $mid = j$, $max\_sp_h = max\_sp_j = max\_sp_l = 3$, and $max\_sp_q = max\_sp_k = 2$. Thus, node q or k is selected as the seed node in step (2-d) of Method 3. In other words, step (2-d) of Method 3 attempts to find the node at the graph-theoretic center of a conflict graph.

4 Computational Experiments

We performed computational experiments to compare the three proposed algorithms with respect to running time and the four adjustment quality measures described in Sect. 2.

In our experiments, we created more than a thousand 50-node graphs. The nodes existed only inside a rectangular region R whose size was 500×500. Each node was assigned an individually-determined height in the range 20–40 and an individually-determined width in the range 30–70.

The FTA and the proposed algorithms are primarily intended to adjust the node positions of a graph. Therefore, we did not use graphs with a large overlap area; i.e., we did not use graphs with an overlap of nodes i and j whose area exceeded one-half of i's area or one-half of j's area.

We used the C programming language to implement the algorithms. All computations were performed on an Intel Core 2 Duo E8600 CPU.

The experimental results are shown in Table 1. All results are presented as the averages of 1000 instances.

Table 1 Experimental results

	Method 1	Method 2	Method 3
λ_1	0.524	0.527	0.504
λ_2	0.256	0.257	0.255
λ_3	0.0370	0.0310	0.0640
f_{cost}	510.6	513.4	448.7
Running time [msec]	0.58	0.71	1.01

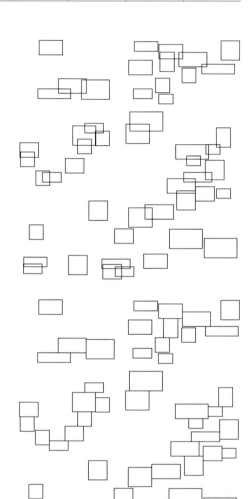

Fig. 7 An example of an original layout

Fig. 8 An example of a layout obtained by Method 1

Fig. 9 An example of a
layout obtained by Method 2

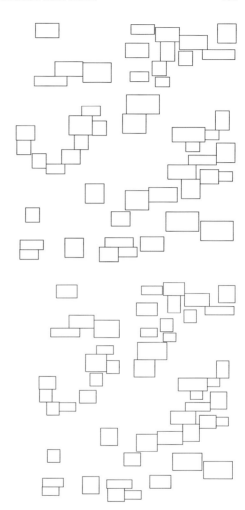

Fig. 10 An example of a
layout obtained by Method 3

The results show that compared to Method 1, Method 2 decreased the layout area without making worse the other adjustment quality measures. In addition, Method 3 decreased the adjusted distances of nodes compared to Method 1; therefore, Method 3 strongly preserves the mental map. However, Method 3 increased the layout area. Note that all the proposed algorithms were sufficiently fast for real-time applications.

We show examples of an original layout and layouts obtained by the proposed algorithms in Figs. 7, 8, 9 and 10.

5 Conclusion

In this study, we have considered removing the node overlaps problem. We have proposed three heuristic algorithms for graphs with several tens of nodes; two of the proposed algorithms are tailored for particular purposes. The experimental results show that the proposed algorithms are effective and fast. It is a future work to compare our algorithms with existing other methods.

References

1. Eades, P.: A heuristic for graph drawing. Congressus Numerantium **42**, 149–160 (1984)
2. Kamada, T., Kawai, S.: An algorithm for drawing general undirected graphs. Inf. Process. Lett. **31**, 7–15 (1989)
3. Fruchterman, T., Reingold, E.: Graph drawing by force-directed placement. Softw. Pract. Exp. **21**(11), 1129–1164 (1991)
4. Sumi, K., Tanaka, H., Ebara, H., Nakano, H.: Performance evaluations of graph drawing algorithms (in Japanese). IEICE Trans. Fund. Electron. Commun. Comput. Sci. J79-A(3), 680-686 (1996)
5. Kawanishi, K., Masuda, S., Yamaguchi, K.: An improvement of Eades' graph drawing algorithm with two kinds of desirable distances (in Japanese). IEICE Trans. Fund. Electron. Commun. Comput. Sci. J83-A(9), 1117-1121 (2000)
6. Misue, K., Eades, P., Lai, W., Sugiyama, K.: Layout adjustment and the mental map. J. Vis. Lang. Comput. **6**, 183–210 (1995)
7. Hayashi, K., Inoue, M., Masuzawa, T., Fujiwara, H.: A layout adjustment problem for disjoint rectangles preserving orthogonal order (in Japanese). IEICE Trans. Fund. Electron. Commun. Comput. Sci. J82-D-I(6), 679-690 (1999)
8. Huang, X., Lai, W., Sajeev, A.S.M., Gao, J.: A new algorithm for removing node overlapping in graph visualization. Inf. Sci. **177**, 2821–2844 (2007)
9. Lyons, K.A., Meijer, H., Rappaport, D.: Algorithms for cluster busting in anchored graph drawing. J. Graph Alg. Appl. **2**(1), 1–24 (1998)
10. Gansner, E.R., Hu, Y.: Efficient node overlap removal using proximity stress model. In: Proceedings of the 16th International Symposium on Graph Drawing (GD 2008). Lecture Notes in Computer Science, vol. 5417, pp. 206-217. Springer, Berlin (2009)
11. Marriott, K., Stuckey, P., Tam, V., He, W.: Removing node overlapping in graph layout using constrained optimization. Constraints **8**(2), 143–171 (2003)
12. T. Dwyer, K. Marriott, P.J. Stuckey: Fast node overlap removal. In: Proceedings of the 13th International Symposium on Graph Drawing (GD 2005). Lecture Notes in Computer Science, vol. 3843, pp. 153-164. Springer, Berlin (2005)
13. Lai, W., Eades, P.: Removing edge-node intersections in drawings of graphs. Inf. Process. Lett. **81**, 105–110 (2002)
14. Abe, N., Masuda, S., Yamaguchi, K.: An algorithm for finding a graph drawing with all vertex labels (in Japanese). IEICE Trans. Fund. Electron. Commun. Comput. Sci J95-A(8), 669-682 (2012)

Significant Frequency Range of Brain Wave Signals for Authentication

Preecha Tangkraingkij

Abstract This study discusses a new biometric system using brain wave signals (EEG). The frequency range of EEG signals is 0–100 Hz, which is categorized into five groups according to their frequency (Delta, Theta, Alpha, Beta, Gamma), however it is noted that all frequency range can degrade in accuracy and recognition speed. The purpose of this study is to explore which frequency range of brain wave signals can be utilized for authentication. In this study, 1,000 data points of EEG signal in group of four channels, F4, P4, C4, and O2 are explored. The practical technique, Independent Component Analysis (ICA) by SOBIRO algorithm is considered clean and separates the individual signals from noise using the technique of supervised neural network for authenticating 20 subjects. From five frequency ranges of EEG signals, it is shown that the best frequency range for the authentication is Delta, which can authenticate 20 subjects within 100 % accuracy.

Keywords Electroencephalogram · Biometric · Authentication · Independent component analysis · Neural network

1 Introduction

Biometrics systems have been used by humans for thousands of years to recognize each other. The term "biometrics" is derived from the Greek words "bio", which means life, and "metric", which means to measure. Biometrics has grown to become an interesting topic in recent years with respect to computer and network security. Biometrics characteristics can be divided into two main classes:

The first is physiological biometrics which relates to the shape of the different parts of body [1], such as: DNA, Ear shape, Face recognition, Fingerprints, Hand and finger geometry, Infrared thermogram, Iris recognition, and Retina. The second

P. Tangkraingkij (✉)
Department of Applied Computer Science, School of Information Technology,
Sripatum University, Bangkok, Thailand
e-mail: pree777@hotmail.com

© Springer International Publishing Switzerland 2016   103
R. Lee (ed.), *Software Engineering, Artificial Intelligence, Networking
and Parallel/Distributed Computing 2015*, Studies in Computational Intelligence 612,
DOI 10.1007/978-3-319-23509-7_8

is behavioral biometrics. This type of biometrics relates to the behaviour of a human such as: Gait, Signature and Typing rhythm (Keystroke).

Biometrics using brain signals have become interesting in research toward using EEG as a biometric measure because the brain is the most complex biological structure known to man and its wave signals are very difficult to mimic or steal. Many techniques such as electroencephalography (EEG), function magnetic resonance imaging (fMRI), magnetoencephalography (MEG), and positron emission tomography (PET) have been utilized for study of the brain. Each technique has its own strengths and weaknesses.

This study proposes the use of EEG because it has a desirable property for excellent temporal resolution, is relatively tolerant of subject movement, and its related hardware costs are lower than other techniques. It has been shown in previous studies that EEG is unique and can be used for biometric identification and authentication [2–9]. Tangkraingkij et al. [10–12] reported the preliminary results of personal authentication of EEG signals by using independent component analysis (ICA) with neural classifier in which EEG signals were used across all frequency ranges of 1,000 data points of 20 subjects, four channels (F4, P4, C4, O2), with the accuracy at 98.51 %. However, all frequency ranges of brain signals can degrade recognition speed and accuracy. In this study, EEG was used to analyze basic brain signals five frequency ranges (Delta, Theta, Alpha, Beta, and Gamma) to the best frequency range for the authentication. The rest of this study is organized as follows. Section 2 summarizes the relevant backgrounds. Section 3 discusses methodology and experimental process. Section 4 is the results and discussions. Section 5 concludes the study.

2 Related Backgrounds

2.1 Electroencephalography (EEG)

EEG is the measurement of electrical activity produced by the brain as recorded from electrodes placed on the scalp. The strength of each signal is considered rather low and the signal measured from any location of scalp can be interfered with by signals from other locations due to the activities of the brain such as eye tracking and EMG (electromyography). In addition, the noise in EEG may be created by the surrounding large electrical potentials from the environment. Brain waves are categorized into five basic groups according to their frequencies as follows: (1) Delta (1–4 Hz), (2) Theta (4–8 Hz), (3) Alpha (8–12 Hz), (4) Beta (12–30 Hz), and (5) Gamma (30–100 Hz). EEG is the most valuable diagnosis of epilepsy. It is also used to help predict a person's chance of recovery after a change in consciousness. The most advanced form of EEG usages is applied in basic Brain Computer Interface (BCI), neuroscience, cognitive science research and statistical signal processing.

2.2 Independent Component Analysis (ICA)

Electroencephalography signal is intrinsically a mixture of the other signals, with such effects as delays, reverberations, and non-linear distortions [13]. It is assumed that the EEG signals from the electrodes on the scalp picks up brain sources and non-brain sources related to movements of eyes and muscles. ICA is a member of a class of blind source separation (BSS). The aim of source separation is to recover original signals from known observations where each observation is an unknown mixture of the original signals. The objective of ICA is to clean and separate the individual signals from different areas of the brain.

2.3 Neural Network Classification Concept

The separated signals from the ICA process cannot be directly used to authenticate a person. The relevant features must be extracted from these signals and the problem of authenticating a person is transformed into a classification problem.

Neural network is a process paradigm that mimics the structures and functions of the human nervous system. Pattern recognition is an important application which can be implemented using a feed-forward neural network that has been trained accordingly. During the training, the network learns to associate output with input patterns. When the network is used, it authenticates the input pattern and tries to output the associated output pattern similar to the way the human brain works.

3 Methodology and Experiments

This research examines brain wave signals by sampling the brain wave signals of 20 subjects and putting them through the independent component analysis to isolate the newly created signal in order to derive the original brain signal. Such a signal can be divided into five intervals, based on the signal frequency. These intervals can be developed further to measure the effectiveness in authenticating a person by using the supervised neural network and to identify which interval of the brain signal is more important than the others. Figure 1 depicts a diagrammatic process block visualization of this study which consists of four main procedures:

3.1 Collecting Brain Wave Signals

Electroencephalography signals were collected from 20 subjects (8 men and 12 women) from the Chulalongkorn Hospital in Bangkok. The age range of the subjects

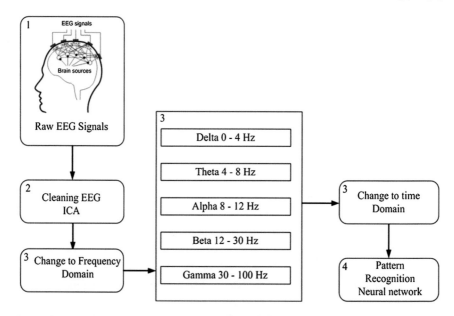

Fig. 1 The experimental diagram process consisting of four main procedures

is between 12 and 40 years. The EEG signals were recorded from 16 electrode channels attached to the scalp while each subject was motionless during the EEG recording experiment and no task was performed. They were not allowed to talk or move during this period. According to 10–20 system, the following locations on the scalp are considered: FP1, F7, T3, T5, FP2, F8, T4, T6, F3, C3, P3, O1, F4, C4, P4, and O2. Figure 2 shows these locations on the scalp. The EEG amplifier was Grass model 8 plus. Our recording sessions used mono-polar montage with reference at the mastoid

Fig. 2 The locations of electrode placements on the scalp using 10–20 system

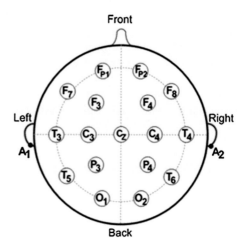

area A1 and A2. The sampling rate was 200 Hz. EEG data were notch filtered at 60 Hz digitized by BMSI board using Stellate Harmony EEG software exported as EDF (European Data Format). For each subject in each trial, 3,000 data points (15 s recording) were simultaneously collected from each of 16 channels.

3.2 Cleaning EEG Signal by ICA

The purpose of this step is to use ICA in isolating the good signal from disturbing signal which has been originally collected. Tangkraingkij et al. [12] found that ICA yielding a good result in authenticating an individual is SOBIRO. Therefore, in this experiment SOBIRO is selected to isolate the good signals by starting with the raw brain signal which has a wavelength of 3,000 data points in all 16 channels and put them through SOBIRO algorithm implemented in ICALAB [14]. Two parameters used in SOBIRO algorithm were set as follows, number of time-delayed covariance matrices was set to 100 and ordering was set to none. The brain waves were re-divided into the five frequency ranges in the next step.

3.3 Dividing the Brain Signal into Five Frequency Ranges

In the past experiment, Tangkraingkij et al. [12] found that the signal which gets the best result is the group of four channel signals. These groups of four channels are F4, P4, C4, and O2. Therefore, this experiment picked these channels to be the sampling model. The purpose of this step is divide brain wave into the five frequency ranges for testing to determine which one is the best for the authentication. It will be divided into three parts for this experiment:

(1) *Converting Time Domain into Frequency Domain*
The experiment starts by selecting the brain signal at channel F4, P4, C4, and O2 which passed the EEG signal cleaning process by ICA. In this experiment, Time Domain is converted into Frequency Domain by using FFT process. The example of the brain wave in Time Domain and Frequency Domain is illustrated in Fig. 3.

(2) *Dividing the brain signal into 5 wave frequency ranges*
After converting the brain wave into frequencies, it will further be divided into 5 intervals as follows: (1) Frequency at 0–4 Hz (Delta wave), (2) Frequency at 4–8 Hz (Theta wave), (3) Frequency at 8–12 Hz (Alpha wave), (4) Frequency at 12–30 Hz (Beta wave), and (5) Frequency at 30–100 Hz (Gamma wave)

(3) *The process of converting Frequency Domain into Time Domain*
Because the pattern recognition process uses the time domain information, it is therefore necessary to convert the brain signal in each frequency range into the time domain by using the reverse FFT process. The brain signal of five frequency ranges

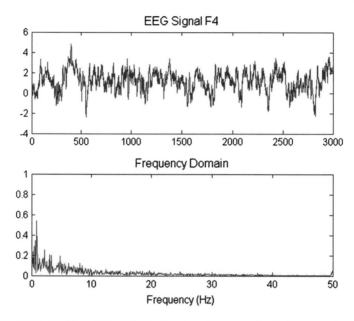

Fig. 3 The EEG signal from F4 location in time domain compared with frequency domain

in time domain will be re-formatted into the 1,000 data points for testing in next step. Figure 4 illustrates Delta, Theta, Alpha, Beta, and Gamma of brain signals in time domain.

3.4 Comparing Each Range of EEG Signals by Neural Network

The purpose of this section is to take the brain wave in each frequency and to compare it using the concept of pattern recognition of the neutral network. From the past experiment, Tangkraingkij et al. [12] used the brain signals of all combined frequencies and found that the frequencies resulting in the most accurate results are F4, P4, C4, and O2 by using a sampling of 1,000 data points of 20 subjects. The accuracy was at 98.51 %

In order to compare the past experiment result, it is necessary to control the position and the wavelength to be equal to compare the results. The sampling data will be the channels F4, P4, C4, and O2 with length of 1,000 data points. This experiment will compare the results from 20 subjects.

Four experiments were conducted:

Experiment1: This experiment compared one frequency range to compare the ability to authenticate 20 subjects. These frequencies are Delta, Theta, Alpha, Beta and Gamma.

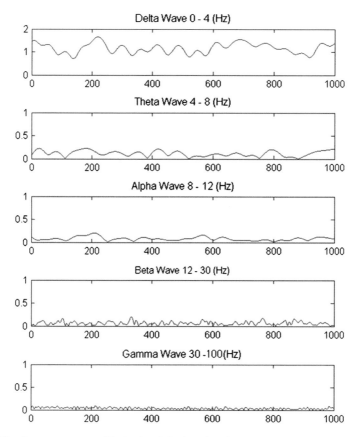

Fig. 4 Five frequency ranges of brain signals in time domain

Experiment 2: This experiment combined the two-frequency range by pairing up the frequencies. The pairing frequencies are Delta-Theta, Delta-Alpha, Delta-Beta, Delta-Gamma, Theta-Alpha, Theta-Beta, Theta- Gamma, Alpha-Beta, Alpha-Gamma, and Beta-Gamma.

Experiment 3: This experiment combined the three frequency range by combing the frequencies. These frequencies are Delta-Theta-Alpha, Delta-Theta-Beta, Delta-Theta-Gamma, Delta-Alpha-Beta, Delta-Alpha-Gamma, Delta-Beta-Gamma, Theta-Alpha-Beta, Theta-Alpha-Gamma, Theta-Beta-Gamma, and Alpha-Beta-Gamma.

Experiment 4: This experiment combined the four-frequency range by combing the frequencies. These five frequencies groups are Delta-Theta-Alpha-Beta, Delta-Theta-Alpha-Gamma, Delta-Theta-Beta-Gamma, Delta- Alpha-Beta-Gamma, and Theta-Alpha-Beta-Gamma.

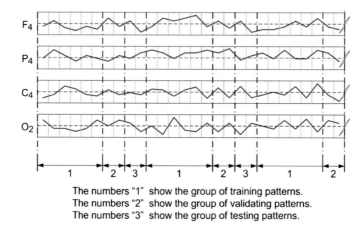

The numbers "1" show the group of training patterns.
The numbers "2" show the group of validating patterns.
The numbers "3" show the group of testing patterns.

Fig. 5 Pattern of data partitioning to form the training, validation, and testing groups

A 3-layer feed-forward neural network with 20 hidden neurons, and 20 output neurons was deployed. Hyperbolic tangent was used as the kernel function and activation function. All training patterns were learnt by a multi-layer perceptron with scaled conjugate gradient backpropagation learning rule. Since four channels were simultaneously considered in this process, each input pattern, including training, validating, and testing consisted of 4 elements. In each sample set, every sequence of 10 data points were grouped as follows: (1) the first six data points were grouped as training patterns, (2) the next two data points were grouped as validating patterns, and (3) the last two data points were grouped as testing patterns. The objective of partitioning data was to get a good representative of the sample points for each data division.

Figure 5 illustrates how training, validating, and testing patterns were grouped from the signals of channels F4, P4, C4, and O2. Numbers 1, 2, and 3 denote the training, validating, testing patterns, respectively. Based on this grouping scheme, it can be seen that 60 % of data points are for training, 20 % are for validating, and another 20 % are for testing.

4 Results and Discussions

In this study, the findings show the effectiveness in using the five brain signal intervals. Frequency 0–4 Hz can correctly authenticate 20 subjects with an accuracy of 100 %. Frequency 4–8 Hz can authenticate the individuals at 36.380 %. Frequency 8–12 Hz can authenticate the individuals at 38.260 %. Frequency 12–30 Hz can authenticate individual at 40.470 %. Frequencies 30–100 Hz can authenticate the individuals at 52.535 % as illustrated in Table 1.

Table 1 The percentage of accuracy for 20 subjects tested with 4 channel combination (F4, P4, C4, O2), SOBIRO, 1,000 data points for 1 frequency range in experiment 1

Wave ranges	Accuracy percentage
Delta	100.000
Theta	36.380
Alpha	38.260
Beta	40.470
Gamma	52.535

Table 2 illustrates the experiment result by combining the two frequencies in the five-frequency range. The Delta Group (Delta-Theta, Delta-Alpha, Delta-Beta, and Delta-Gamma) proves to have the accuracy rate of 99.975–100 %. The accuracy in the non-Delta group (Theta-alpha, Theta-Beta, Theta-Gamma, Alpha-Beta, Alpha-Gamma, and Beta-Gamma) is 31.070–46.130 %. Comparing the non-Delta Group to Delta Group proves the result to be much better for the Delta Group.

Combing the three-frequency range proves the Delta Group (Delta-Theta-Alpha, Delta-Theta-Beta, Delta-Theta-Gamma, Delta-Alpha-Beta, Delta-Alpha-Gamma, and Delta-Beta-Gamma) to have the accuracy of 95.000–100 % and the non-Delta Group (Theta-Alpha-Beta, Theta-Alpha-Gamma, Theta-Beta-Gamma, and Alpha-Beta-Gamma) is proven to have the accuracy of only 32.020–42.375 %. The detail is illustrated in Table 3.

Table 4 illustrates the four-frequency experiment. The result proves to be similar to the two-frequency group and the three-frequency group. The accuracy in the Delta Group (Delta-Theta-Alpha-Beta, Delta-Theta-Alpha-Gamma, Delta-Theta-Beta-Gamma, and Delta-Alpha-Beta-Gamma) is 95.000–99.995 % while the accuracy in the non-Delta Group (Theta-Alpha-Beta-Gamma) is 37.955 %. From the experiment, it can be seen that the Delta component of the brain signal is an important aspect in authenticating. If such a particular Delta signal is used in the 20 subjects, the

Table 2 The percentage of accuracy for 20 subjects tested with 4 channel combination (F4, P4, C4, O2), SOBIRO, 1,000 data points for 2 frequency ranges combination in experiment 2

Wave ranges		Accuracy percentage
Delta	Theta	100.000
Delta	Alpha	100.000
Delta	Beta	100.000
Delta	Gamma	99.975
Theta	Alpha	31.070
Theta	Beta	32.670
Theta	Gamma	41.360
Alpha	Beta	35.530
Alpha	Gamma	43.500
Beta	Gamma	46.130

Table 3 The percentage of accuracy for 20 subjects tested with 4 channel combination (F4, P4, C4, O2), SOBIRO, 1,000 data points for 3 frequency ranges combination in experiment 3

Wave ranges			Accuracy percentage
Delta	Theta	Alpha	100.000
Delta	Theta	Beta	95.000
Delta	Theta	Gamma	100.000
Delta	Alpha	Beta	95.000
Delta	Alpha	Gamma	100.000
Delta	Beta	Gamma	100.000
Theta	Alpha	Beta	32.020
Theta	Alpha	Gamma	38.680
Theta	Beta	Gamma	39.710
Alpha	Beta	Gamma	42.375

Table 4 The percentage of accuracy for 20 subjects tested with 4 channel combination (F4, P4, C4, O2), SOBIRO, 1,000 data points for 4 frequency ranges combination in experiment 4

Wave ranges				Accuracy percentage
Delta	Theta	Alpha	Beta	95.000
Delta	Theta	Alpha	Gamma	99.995
Delta	Theta	Beta	Gamma	99.990
Delta	Alpha	Beta	Gamma	99.995
Theta	Alpha	Beta	Gamma	37.955

accuracy is proved to 100 %. If the Delta component is combined in a two-frequency experiment, the result still proves to be good. However, if the Delta component is combined in the three-frequency experiment or more, the result proves to be less accurate.

Delta wave frequency is 0–4 Hz. This wave range tends to be the highest in amplitude and the slowest waves. The amplitude of Delta wave is higher than the others, causing a variety of grouping by neural network. Delta wave range has been used the group of position F4, P4, C4, and O2 in this experiment. It will be interesting to look at the other factors such as the different position for Delta wave range, different individual group with more subjects and different timing in the next research.

5 Conclusion

This research studied the brain wave signals (EEG) having the frequency of 0–100. The frequency was divided into intervals to find the accuracy in the authentication subjects in each range. The experiment combined the different frequencies to form the different test case. The brain signals 1,000 data points from position F4, P4, C4, and O2 were used. The experiment used SOBIRO algorithm to bring back the original signal then a supervised neural network was used to test the accuracy of the authentication for 20 subjects. The experiment combined the different frequencies to form the different test cases from one frequency range, two frequency ranges, three frequency ranges, and four frequency ranges. The results show that Delta range of brain wave signals is actually significant for the authentication.

Acknowledgments Thank you to Tayard Desudchit, MD., Neurology & Clinical Neurophysiology, Chulalongkorn Hospital, Thailand who generated the raw EEG data. This research is supported by grants from Sripatum University.

References

1. Jain, A.K., Ross, A., Prabhakar, S.: An introduction to biometric recognition. IEEE Trans. Circuits Syst. Video Technol. **14**(1), 4–20 (2004)
2. Paranjape, R.B., Mahovsky, J., Benedicenti, L., Koles, Z.: The electroencephalogram as a biometrics. Proc. Can. Conf. Electr. Comput. Eng. **2**, 1363–1366 (2001)
3. Poulos, M., Rangoussi, M., Alexandris, N., Evangelou, A.: A on the use of EEG features towards person identification via neural networks. Med. Inf. Internet Med. **26**(1), 35–48 (2001)
4. Poulos, M., Rangoussi, M., Alexandris, N., Evangelou, A.: Person identification from the EEG using nonlinear signal classification. Methods Inf. Med. **41**(1), 64–75 (2002)
5. Palaniappan, R, Ravi, K.V.R.: A new method to identify individuals using signals from the brain. In: Proceedings of fourth international conference information communication and signal processing, pp. 15–18 (2003)
6. Palaniappan, R., Mandic, D.P.: Biometrics from brain electrical activity: a machine learning approach. IEEE Trans. Pattern Anal. Mach. Intell. **29**, 738–742 (2007)
7. Palaniappan, R.: Method of identifying individuals using VEP signals and neural network. IEEE Proc. Sci. Meas. Technol. **151**(1), 16–20 (2004)
8. Palaniappan, R., Mandic, D.P.: EEG based biometric framework for automatic identity verification. VLSI Signal Process **2**(2), 243–250 (2007)
9. Marcel, S., Millan, J.: Person authentication using brainwaves (EEG) and maximum a posteriori model adaptation. IEEE Trans. Pattern Anal. Mach. Intell. **29**(4), 743–752 (2007)
10. Tangkraingkij, P., Lursinsap, C., Sanguansintukul, S., Desudchit, T.: Selecting relevant EEG signal locations for personal identification problem using ICA and neural network. In: Eighth IEEE/ACIS International Conference on Computer and Information Science (ICIS 2009), pp. 616–621 (2009)
11. Tangkraingkij, P., Lursinsap, C., Sanguansintukul, S., Desudchit, T.: Personal identification by EEG using ICA and neural network. In: Computational Science and Its Applications (ICCSA 2010). Lecture Notes in Computer Science, vol. 6018, pp. 419–430 (2010)
12. Tangkraingkij, P., Lursinsap, C., Sanguansintukul, S., Desudchit, T.: Insider and outsider person authentication with minimum number of brain wave signals by neural and homogeneous identity filtering. Neural Comput. Appl. **22**(1 Supplement), 463–476 (2013)
13. Cichocki, A.: Blind signal processing methods for analyzing multichannel brain signals. Int. J. Bioelectromagn. **6**, 1 (2004)
14. Cichocki, A., Amari, S., Siwek, K., Tanaka T., et al.: ICALAB toolboxes. http://www.bsp.brain.riken.jp/ICALAB

Simple Models Characterizing the Cell Dwell Time with a Log-Normal Distribution

Naoshi Sakamoto

Abstract For designing a wireless network, the distribution of the cell dwell time of nodes is important. It was reported that the distribution of the cell dwell time is approximated by a log-normal distribution. Thus, in this paper, we present two simple models in order to estimate the probabilistic distribution of the cell dwell time. One is the model where a node moves straightforward and goes across a cell where the velocity of a node is given by a probabilistic density function. Another is the model of a random walk, named a "tipsy random walk," where a node moves with constant velocity and turns gently. We show that the probabilistic distribution of the cell dwell time of each model can be approximated by a log-normal distribution.

1 Introduction

As mobile terminals spread, it becomes increasingly important to plan and design to establish base stations of wireless LAN. In wireless LAN, a base station basically communicates only one terminal in a moment, then it repeats to communicate every terminal in the communication area. Thus, the communication capacity of the base station is divided by the terminals in the area. Therefore, it is important to establish base stations appropriately for the density of terminals.

In the problem about establishment of base terminals for wireless LAN, it is generally assumed that the performance of each base station is the same, and only one base station takes charge of an area. Then, we can assume that the shape of each area is the same. We call an area of which a base station takes charge a "cell". On the other hand, we naturally consider that mobile terminals can move. Thus, the number of terminals in a cell may be changed every moment. Therefore, the use time for a base station of each terminal is important for the problem about establishment of base stations.

N. Sakamoto (✉)
Tokyo Denki University, Tokyo, Japan
e-mail: sakamoto@c.dendai.ac.jp

© Springer International Publishing Switzerland 2016   115
R. Lee (ed.), *Software Engineering, Artificial Intelligence, Networking
and Parallel/Distributed Computing 2015*, Studies in Computational Intelligence 612,
DOI 10.1007/978-3-319-23509-7_9

Recently, various kinds of investigation for mobile communication have been made. They were investigated by methods of actual surveying, theoretical analyses with an adequate model, and using a random walk with several assumptions. Specially, the author is interested in the paper "Vehicle mobility characterization based on measurement and its application to cellular communication systems," written by Kobayashi et al. [8]. According to this paper, the distribution of the cell dwell time of actual taxis can be approximated by a log-normal distribution. Thus, in this paper, we propose two models of which the distribution of the cell dwell can be approximated by a log-normal distribution.

In former works, the analysis for the channel holding time and the hand-off rate have been regarded as important. In a wireless LAN, a call is also generated and held as well as a wired LAN. Moreover, a hand-off happens when a node reaches the boundary of a cell while it is holding a call, in a wireless LAN. It is the fact that the establishment of base stations depends on the distribution of participants indeed. On the other hand, we can consider that for a node, generating and holding a call are independent against the cells. Therefore, we can consider that we will be able to compute the channel holding time and the hand-off rate by using the cell dwell time and the assumptions of a call. Actually, Zonoozi and Dassanayake computed the channel holding time by using the cell dwell time and the call duration [9].

On the other hand, the property of a random walk has been studied variously. In the former works, in order to conform to mobile communication, the assumption of the random walk of each study usually contains that a node appears at arbitrary location, and also turns to arbitrary angle [4, 7, 9]. However, the author doubts whether these assumptions are appropriate for the motion of actual vehicle or human. Kobayashi et al. actually surveyed the time that taxis go across the cell that is considered virtually by them. In order to argue the model of motion of vehicles or humans, in the author's opinion, it should not be assumed that a node appears at arbitrary location, but the time from entering to exiting the node must be surveyed.

The organization of this paper is as follows.

In Sect. 2, for given a probabilistic density function(PDF) of the velocity of mobile nodes, we induce the PDF of the cell dwell time. Moreover, we see that the distribution of the cell dwell time can be approximated by a log-normal distribution when the velocity of mobile nodes obeys a uniform distribution, and mobile nodes go across a circle.

In Sect. 3, we propose a notion of a "tipsy random walk" that enables to control its degree of the straightness by a parameter. We show that for a round cell, the distribution of the period from the time that a node enters to the time that it exits can also be approximated by a log-normal distribution.

Finally, in Sect. 4, we conclude the results.

2 Transformation of a Probabilistic Distribution Function of Velocity

In this section, we induce formulas of the PDF of the cell dwell time by transforming the PDF of the velocity of mobile nodes. Moreover, by assuming the distribution of the velocity of a mobile node as a uniform distribution, we show that the obtained PDFs of the cell dwell time can be approximated by a log-normal distribution.

2.1 Related Works

In the former works, not only the cell dwell time, but also the channel holding time and the hand-off rate have been investigated.

Hong and Rappaport theoretically studied the channel holding time with assumption that the velocity is uniformly distributed from 0 to V_{MAX} and the direction is also uniformly distributed from 0 to 2π independently [6]. Note that they assumed that the happen and the endurance of a call is exponentially distributed.

Cho et al. induced and evaluated the formula of the cell resident time and the hand-off rate from the PDF of the displacement and the velocity of mobile nodes [3]. First, they proposed the integral transformation to the cumulative distribution function(CDF) of the cell dwell time. Then, they induced the formula of the CDF of the hand-off rate with the assumption that the velocity is uniformly distributed from 0 to V_{MAX} and the direction is also uniformly distributed from 0 to 2π independently where the cell is the round shape. Finally, they showed that their formula approximates the function of Hong and Rappaport [6] by using the numerical integration method.

With respect to the analysis of Cho et al. [3], Boche and Jugl proposed a model for wireless communication system and induced the distribution of the cell dwell time for the velocity distribution theoretically for the system [1]. Specially, they deeply investigated the case when the velocity may be 0.

2.2 Assumption

We fix the size of a cell to l, as Fig. 1. We will induce the PDF of the cell dwell time of mobile nodes that go across a cell from a given PDF of the velocity. In this section we assume the following conditions:

1. the velocity of each node is not changed in the cell, and
2. every node goes across the cell straightforward.

Fig. 1 Shapes of a cell for
two-dimensional area

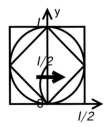

Let $m(v)$ denote the PDF of the velocity of nodes. That is, the probability that the velocity within which a node moves in the cell is less than V_0 is defined as the following formula:

$$\Pr[v \leq V_0] = \int_0^{V_0} m(v)dv.$$

2.3 One-Dimensional Path

We induce the PDF $c_1(t)$ of the cell dwell time where the cell is an one-dimensional path with length l.

Theorem 1 *The PDF $c_1(t)$ of the cell dwell time where the cell is a one-dimensional path, is (1).*

$$c_1(t) = \frac{l}{t^2} m\left(\frac{l}{t}\right). \tag{1}$$

Proof By the assumption, once a velocity is determined for each node, a node holds the velocity as a constant. Thus, for the velocity V, the cell dwell time is $T = l/V$. Then, we have the probability when the cell dwell time is in the period from a to b as following:

$$\begin{aligned}
\Pr[a \leq T \leq b] &= \Pr[a \leq l/V \leq b] \\
&= \Pr[l/b \leq V \leq l/a] \\
&= \int_{l/b}^{l/a} m(v)dv \\
&= \int_a^b \frac{l}{u^2} m\left(\frac{l}{u}\right) du. \qquad \Box
\end{aligned}$$

Corollary 1 *Let $\bar{m}$ be the PDF of a uniform distribution from v_0 to v_1. That is, $\bar{m}$ is given by (2).*

$$\bar{m}(v) = \begin{cases} 0 & \text{if } v < v_0, \\ \frac{l}{v_1 - v_0} & \text{if } v_0 \le v \le v_1, \\ 0 & \text{o.w.} \end{cases} \tag{2}$$

Then, we can obtain the PDF $\bar{c}_1(t)$ of the cell dwell time for $\bar{m}$ as following:

$$\bar{c}_1(t) = \begin{cases} 0 & \text{if } t < \frac{l}{v_1}, \\ \frac{l}{(v_1 - v_0)t^2} & \text{if } \frac{l}{v_1} \le t \le \frac{l}{v_0}, \\ 0 & \text{o.w.} \end{cases}$$

Moreover, we also obtain the CDF $\bar{C}_1(t)$ as following:

$$\bar{C}_1(t) = \begin{cases} 0 & \text{if } t < \frac{l}{v_1}, \\ \frac{l}{v_1 - v_0} \left(\frac{v_1}{l} - \frac{1}{t} \right) & \text{if } \frac{l}{v_1} \le t \le \frac{l}{v_0}, \\ 1 & \text{o.w.} \end{cases}$$

2.4 Two-Dimensional Area

We calculate the PDF $c(t)$ of the cell dwell time where the cell is a two-dimensional area. At first, we assume that mobile nodes go across the cell straightforward with a constant velocity.

We have to discuss the shape of a cell. We can ignore the angle of incidence when the shape of the cell is round. However, circles can not tile a plane. Thus, the shape of a cell is used to be assumed to be a square or a hexagon. On the other hand, the boundary of an actual cell is not uniquely determined. It might always depend on the shape of the ground and the positional relationship between the base stations. Thus, we choose a circle with radius $l/2$, a inscribed square and a circumscribed square as the shape of a cell (Fig. 1). Moreover, let the incidence angle be horizontal. Notice that at a cell, since the cell dwell time when a node goes across the upper half is the same as the dwell time when a node goes across the lower half, we only consider the lower half of a cell. That is, we only consider that y is in the region from 0 to $l/2$. Moreover, the position of the y-axis that a node goes across is assumed to be uniformly distributed between 0 and $l/2$. Then, we have the PDF of y as the following (3):

$$f(y) = \begin{cases} 0 & \text{if } y < 0, \\ \frac{2}{l} & \text{if } 0 \le y \le \frac{l}{2}, \\ 0 & \text{o.w.} \end{cases} \tag{3}$$

2.4.1 Circumscribed Square

For the circumscribed square, since the cell dwell time does not depend on y, the PDF is the same as the PDF for an one-dimensional path.

2.4.2 Inscribed Square

If a node goes across a cell whose shape is an inscribed square, the cell dwell time T is $2y/V$ where a node goes across the y-axis at the y-coordinate y with velocity V in the circumscribed square $(0 \leq y \leq l/2)$.

Theorem 2 *When the velocity is distributed with a PDF $m(v)$, we have the PDF $c_d(t)$ of the cell dwell time where the cell is the inscribed square of the circle of radius $l/2$ as the following (4):*

$$c_d(t) = \frac{4}{lt^2} \int_0^{l/2} m\left(\frac{2s}{t}\right) s \, ds. \tag{4}$$

Proof We transform the formula of the probability that the cell dwell time is in the period from a to b as the following:

$$\Pr[a \leq T \leq b] = \int_{-\infty}^{\infty} \Pr[a \leq T \leq b | y = s] f(s) ds$$

$$= \int_{-\infty}^{\infty} \Pr\left[\frac{2s}{b} \leq v \leq \frac{2s}{a}\right] f(s) ds$$

$$= \int_{-\infty}^{\infty} \int_{2s/b}^{2s/a} m(v) f(s) dv ds. \tag{5}$$

We apply Fubini's theorem. Moreover, we compute the integration by substitution. Let $u = 2s/v$, then this yields $dv = -2s/u^2 du$. Finally, we apply (3). Then, we have the following:

$$= \int_a^b \int_0^{l/2} \frac{4s}{lu^2} m\left(\frac{2s}{u}\right) ds \, du. \qquad \qquad \square$$

Corollary 2 *When the velocity of nodes is uniformly distributed from v_0 to v_1, we have the PDF $\bar{c}_d(t)$ of the cell dwell time where the cell is the inscribed square of the circle of radius $l/2$ as follows:*

$$\bar{c}_d(t) = \begin{cases} \frac{v_1 + v_0}{2l} & \text{if } t < \frac{l}{v_1}, \\ \frac{1}{2l(v_1 - v_0)} \left(\frac{l^2}{t^2} - v_0^2\right) & \text{if } \frac{l}{v_1} \leq t \leq \frac{l}{v_0}, \\ 0 & \text{o.w.} \end{cases}$$

Moreover, we can have the CDF $\bar{C}_d(t)$ as follows:

$$\bar{C}_d(t) = \begin{cases} \frac{v_1+v_0}{2l}t & \text{if } t < \frac{l}{v_1}, \\ \frac{1}{v_1-v_0}\left(v_1 - \frac{v_0^2}{2}\frac{t}{l} - \frac{1}{2}\frac{l}{t}\right) & \text{if } \frac{l}{v_1} \le t \le \frac{l}{v_0}, \\ 1 & \text{o.w.} \end{cases}$$

2.4.3 Circle

Next, we consider that a node goes across the cell whose shape is round and radius is $l/2$.

For any incidence angle, we can easily transform it to horizontal by rotation. Thus, we consider not the upper and the lower sides of the circle, but the lower side of the circle only, and let nodes enter between 0 and $l/2$ horizontally, and the y-coordinate is assumed to be uniformly distributed. On the other hand, the cell dwell time T is $2\sqrt{ly-y^2}/V$.

Theorem 3 *When the velocity is distributed with a PDF $m(v)$, we have the PDF $c_c(t)$ of the cell dwell time where a cell is a circle of radius $l/2$ as the following (6):*

$$c_c(t) = \frac{4}{lt^2}\int_0^{l/2}\sqrt{ls-s^2}\,m\left(\frac{\sqrt{ls-s^2}}{t}\right)ds. \tag{6}$$

Proof As similar to the proof of Theorem 2, we transform the formula of the probability that the cell dwell time is in the period from a to b as following:

$$\begin{aligned} &\Pr[a \le T \le b]\\ &= \int_{-\infty}^{\infty} \Pr[a \le T \le b|y=s]f(s)ds\\ &= \int_{-\infty}^{\infty} \Pr\left[\frac{2\sqrt{ls-s^2}}{b} \le v \le \frac{2\sqrt{ls-s^2}}{a}\right]f(s)ds\\ &= \int_{-\infty}^{\infty} \int_{2\sqrt{ls-s^2}/b}^{2\sqrt{ls-s^2}/a} m(v)f(s)dvds. \end{aligned} \tag{7}$$

We also apply Fubini's theorem. Moreover, we compute the integration by substitution. Let $u = 2\sqrt{ls-s^2}/v$, then this yields $dv = (-2\sqrt{ls-s^2}/u^2)du$. Finally, we apply (3). Then, we have the following:

$$= \int_a^b \frac{4}{lu^2}\int_0^{l/2}\sqrt{ls-s^2}\,m\left(\frac{2\sqrt{ls-s^2}}{u}\right)dsdu. \qquad \square$$

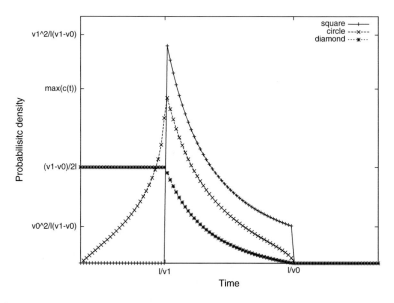

Fig. 2 Probabilistic density functions

Corollary 3 *When the velocity of nodes is uniformly distributed from v_0 to v_1, we have the PDF $\bar{c}_c(t)$ of the cell dwell time where a cell is a circle of radius $l/2$ as follows:*

$$
\bar{c}_c(t) =
\begin{cases}
\frac{l}{t^2}\left(\arcsin(v_1 t/l) - \arcsin(v_0 t/l)\right)/(v_1 - v_0) \\
\quad - \left(\frac{v_1}{lt}\sqrt{l^2 - t^2 v_1^2} - \frac{v_0}{lt}\sqrt{l^2 - t^2 v_0^2}\right)/(v_1 - v_0), & \text{if } t < \frac{l}{v_1}, \\
\frac{l}{t^2}\left(\frac{\pi}{2} - \arcsin(v_0 t/l)\right)/(v_1 - v_0) \\
\quad + \frac{v_0}{lt}\sqrt{l^2 - t^2 v_0^2}/(v_1 - v_0), & \text{if } \frac{l}{v_1} \le t \le \frac{l}{v_0}, \\
0 & o.w.
\end{cases}
$$

We graphically show the PDFs when the velocity is uniformly distributed from v_0 to v_1 in Fig. 2. On the other hand, we graphically show the CDFs on a log-normal probability plotting paper in Fig. 3. The axis of abscissas is log scale. The line of the $\bar{C}_c(t)$ is drawn by using the numerical integration method. According to the Fig. 3, we can see that none of these CDFs exactly corresponds to a log-normal distribution, but we can say that they approximate a log-normal distribution between 10 and 90 %.

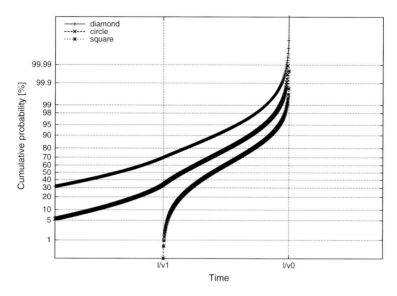

Fig. 3 Cumulative distribution functions on log-normal probability plotting paper

3 Random Walk

Next, we propose a simple notion of a random walk of which the distribution of the cell dwell time approximates a log-normal distribution.

3.1 Related Works

The property of wireless LAN has been investigated by also simulating a random walk.

Chiang et al. analyzed the cell dwell time of a two-dimensional random walk for a hexagon cell by dividing the cell into small hexagons [2].

Zonoozi and Dassanayake consider the following random walk in a hexagon cell [9]:

1. a node starts at arbitrary location in the area,
2. the direction is uniformly chosen from arbitrary angle,
3. the direction is uniformly changed within the predefined limit $\pm\alpha$ for the current direction,
4. the velocity at the beginning is chosen by Gaussian distribution from 0 to 100 [km/h],
5. the velocity is uniformly changed in the range $\pm10\,\%$ of the current velocity.

They evaluated the distribution of the cell residence time obtained by the simulation by using Kolmogorov-Simnov goodness-of-fit test with respect to the best-fit generalized gamma distribution. Moreover, they also studied the distribution of the handover rate and the channel holding time.

Guérin simulated a random walk to analyze the channel holding time and also studied it theoretically [4]. On the simulation of the random walk, he assumed that the shape of a cell is round, the velocity of a node is a constant, and the appearing location and the direction are chosen with a uniform distribution. Moreover, he also assumed that a node determines the time to move straightforward by an exponential distribution, after the time spends, determines the new direction with a uniform distribution, and determines the time to move straightforward again. Moreover, he theoretically analyzed the number of hand-offs and the channel holding time by proposing a model. In the analysis, he assumed that the shape of a cell is a hexagon and the directions that a node is allowed to move are up and down, left and right.

Jabbari et al. studied the channel holding time in a round or a square cell by considering a two-dimensional random walk and a Markov chain as well as our study [7]. They considered that a cell is divided into small square areas and the random walk is assumed that a node is allowed to move to up and down, left and right. Moreover, they assumed that the probability that a node moves to the same direction as the previous is larger. This approach is similar to ours. However, the points of difference between their approach and ours are the distribution of the probability and the evaluation.

3.2 Two-Dimensional Random Walk Without Turning Back

In Sect. 2, in order to approximate the cell dwell time from a given velocity distribution, we have made the strong assumption on the shape of a cell and the condition of the velocity. Actually, the assumption might be held in small cells. On the other hand, the obtained PDF of the cell dwell time does not depend on the size of a cell. By analyzing the result of Kobayashi [8], Hidaka showed that the cell dwell time is approximated by the exponential distribution for a small cell [5]. Therefore, our argument in Sect. 2 may be appropriate to the certain size of a cell only.

Then, we would like to propose a model where the cell dwell time depends on the size of a cell. Moreover, it is ideal that the model enables to be controlled by the size of a cell. As the candidate for this, we propose a random walk that has the special condition.

In the case of former models of a random walk, since a node averagely stays around the starting point with high probability, these models are not appropriate to apply to the problem concerning that a node goes across the cell. Thus, at first, we propose the two-dimensional random walk where a node moves to the up and down, left and right direction without turning back. Then, we measure the cell dwell time for the model by computer simulation. We graphically show the results in Fig. 4. According to the result, nodes might exit from the cell in early time with high probability, regardless

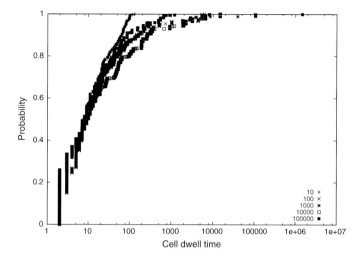

Fig. 4 Cell dwell time of random walk without turning back

of the size of the cell. Figure 4 shows the CDFs of the cell dwell time where the sizes of a cell are 10, 100, 1000, 10,000, and 100,000. The axis of abscissas is log scale. We can see that each cumulative distribution under 90 % is similar to each other, regardless the size of a cell.

We find two consideration. One is that the direction of nodes tends to become random for sufficiently long time. Another is that many nodes do not exit from the opposite boundary of the cell, but return to the side of the start point. Therefore, in this simulation, we find that a random walk without turning back becomes similar to a normal random walk.

3.3 Tipsy Random Walk

Then, we extend the notion of a random walk without turning back. We consider that a node can choose a direction from still only three directions at each time. Moreover, let the directions be narrower. Notice that a random walk where the directions are angle $\pm 90°$ is equivalent to a random walk without turning back. We denote a random walk where the narrower directions are allowed a "tipsy random walk." Then, we observe the property of a tipsy random walk by computer simulation.

First, we can say that the transition model of the direction of a node can be applied to a Markov chain. We denote n as the fraction size of the circumference. When the allowed directions are $\pm 2\pi/n$, we have the transition matrix $P = (p_{ij})$ $(1 \leq i, j \leq n)$ as (8).

$$p_{i,j} = \begin{cases} \frac{1}{3} & \text{if } i = j - 1, j, j + 1 \text{ or } (i, j) = (1, n), \\ 0 & \text{o.w.} \end{cases} \tag{8}$$

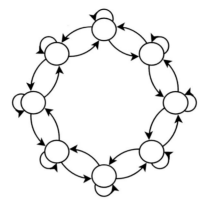

Fig. 5 Transition model of the direction, when $n = 8$

We show a state diagram in Fig. 5, when $n = 8$. For this Markov chain, we calculate the probability of staying each direction for every time. This probability shows the probability which direction a node moves to. It is clear that the state probability of the direction of the start is the greatest, and the state probability of the opposite direction against the start is the smallest. However, we can see that the probability of every direction comes to be uniform. We show the probability of each direction (0, 90, and 180°) in Fig. 6 where $n = 64$. The axis of abscissas is log scale of time. And, the axis of ordinates is the probability rate to assume $1/n$ as 1. Thus, the rate at the direction of the start at time 0 is equal to n. As Fig. 6 shows, the probability of staying every direction always converses to rate 1, after long time.

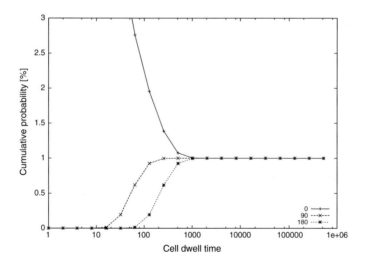

Fig. 6 Probability of staying, when $n = 64$

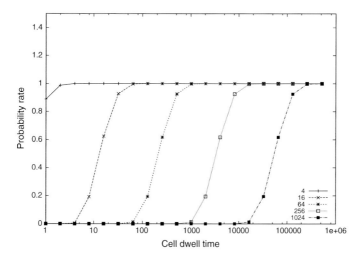

Fig. 7 Probability of staying the opposite direction

With respect to the fraction size, we show the cumulative probability of staying the opposite direction against the start in Fig. 7. According to Fig. 7, we find that we can control the time to converge by the fraction size. That is, we can assume that there exists a function $T(n)$ where nodes move almost straightforward in the time shorter than $T(n)$. And, nodes move randomly in the time longer than $T(n)$.

Then, we measure the distribution of the time that a node exits a cell after the node is put at the boundary of the circle by computer simulation (Fig. 8). We repeat to simulate this by letting the fraction size be 4, 16, 64, 256, and 1024, and the diameter of the circles be for 100 and 10,000, respectively, for 10,000 times. We show the obtained distributions of the cell dwell time in Figs. 9 and 10. The axis of abscissas is log scale of time. And, the axis of ordinates denotes probability with the inverse of the normal distribution axis. We can see that each distribution might not be considered as a straight line, but approximate to the straight line for the region between 20 and 90 %. Thus, we can say that a tipsy random walk can roughly approximate a log-normal distribution.

The fact that the slope is steep implies that the time to exit is in the narrow period. In this case, we can say that most of nodes move straightforward. On the other hand, the fact that the slope is gentle implies that there exist both nodes that exit in short time and nodes that exit in long time. In this case, we can say that most of nodes move randomly.

Fig. 8 Scenario of the simulation

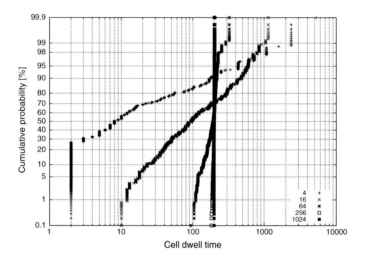

Fig. 9 Cell dwell time where the size of the circle is 100

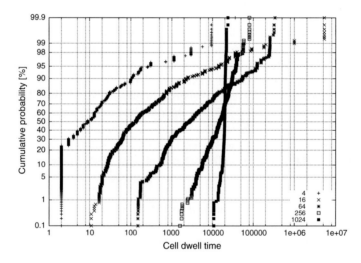

Fig. 10 Cell dwell time where the size of the circle is 10,000

Note that the scales of the axis of abscissas of Figs. 9 and 10 are different. Nevertheless, when we focus on the lines where the fraction size is 64, we can find that the slope is steep when the diameter is equal to 100 but the slope is gentle when the diameter is equal to 10,000. According to Fig. 7, the time to converge is likely greater then 100 and less than 1000. That is, in this case, we can say that nodes move straightforward before time 100 and move randomly after time 10,000 where $n = 64$. Then, we can consider that nodes exit a cell straightforward for the circle of diameter 100, but exit at random location of the boundary for the circle of diameter 10,000. Therefore, we can say that the notion of a tipsy random walk enables to control the degree of the straightness by a parameter.

4 Conclusion

Since it was reported that the actual cell dwell time is approximated by a log-normal distribution [8], we propose two models where the cell dwell time is approximated by a log-normal distribution. One is the model that nodes go across a cell straightforward for a given velocity distribution. In particular, when the velocity is uniformly distributed, the cell dwell time is approximated by a log-normal distribution where the shape of a cell is round. We can say that when the actual mobile nodes are expected to move straightforward, we can estimate the distribution of the cell dwell time by the velocity distribution of the mobile nodes.

On the other hand, we propose another model as a tipsy random walk. This is a random walk that turns gradually. We can say that this is the model where nodes move almost straightforward in short time, and move randomly in long time. Then, we can have the distribution approximated by a log-normal distribution by letting the velocity be a constant, and giving the fraction size for the angle of a turn.

In the future, we would like to reveal the relationship between the distribution of our model and the known other distribution. Specially, since Hidaka et al. reported that the distribution of the cell dwell time in the small cell is approximated to an exponential distribution [5], we have to reveal the relationship between the distribution yielded for the size of a cell and an exponential distribution. Moreover, we would like to estimate time $T(n)$ that classifies the behavior of a tipsy random walk where n is the fraction size.

References

1. Boche, H., Jugl, E.: Dwell time modeling for wireless communication systems and problems of the velocity distribution. Eur. Trans. Telecommun. **13**(3), 269–278 (2002). http://dblp.uni-trier.de/db/journals/ett/ett13.html#BocheJ02
2. Chiang, K.H., Shenoy, N.: A 2-d random-walk mobility model for location-management studies in wireless networks. IEEE Trans. Veh. Technol. **53**(2), 413–424 (2004). http://dblp.uni-trier.de/db/journals/tvt/tvt53.html#ChiangS04
3. Cho, M., Kim, K., Szidarovszky, F., You, Y., Cho, K.: Numerical analysis of the dwell time distribution in mobile cellular communication systems. IEICE Trans. Commun. **E81-B**(4), 715–721 (1998)
4. Guérin, R.: Channel occupancy time distribution in a cellular radio system. IEEE Trans. Veh. Technol. **VT-36**(3), 89–99 (1987)
5. Hidaka, H., Saitoh, K., Shinagawa, N., Kobayashi, T.: Teletraffic characterization of cellular communication for different types of vehicle motion. IEICE Trans. Commun. **E84-B**(3), 558–565 (2001)

6. Hong, D., Rappaport, S.S.: Traffic model and performance analysis for cellular mobile radio telephone systems with prioritized and nonprioritized handoff procedures. IEEE Trans. Veh. Technol. **35**(3), 77–92 (1986). doi:10.1109/T-VT.1986.24076
7. Jabbari, B., Zhou, Y., Hillier, F.: Random walk modeling of mobility in wireless networks. In: 48th IEEE Transaction on Vehicular Technology Conference, 1998. VTC 98., vol. 1, pp. 639–643 (1998). doi:10.1109/VETEC.1998.686653
8. Kobayashi, T., Shinagawa, N., Watanabe, Y.: Vehicle mobility characterization based on measurement and its application to cellular communication systems. IEICE Trans. Commun. **E82-B**(12), 2055–2060 (1999)
9. Zonoozi, M., Dassanayake, P.: User mobility modeling and characterization of mobility patterns. IEEE J. Sel. Areas Commun. **15**(7), 1239–1252 (1997)

A Method of Ridge Detection in Triangular Dissections Generated by Homogeneous Rectangular Dissections

Koichi Anada, Taiyou Kikuchi, Shinji Koka, Youzou Miyadera
and Takeo Yaku

Abstract In order to display 3D terrain map effectively, detections of features on maps are very important. In this paper, we discuss a method for detection of ridges. It is known in the previous work that the steepest ascent method is effective for a ridge detection on terrain map represented by rectangular dissections. We will introduce the steepest ascent method in triangular dissections generated by homogeneous rectangular dissections.

1 Introduction

In order to display 3D terrain map effectively, detections of features on maps are very important. In this paper, we discuss a method for detection of ridges.

In [5], Yokoyama et. al. introduced a method to detect ridges in terrain maps called the steepest descent line method. Steepest descent lines are similar to ones by the drop of water principle (cf. [1]).

K. Anada (✉)
Research Institute for Science and Engineering, Waseda University, Tokyo, Japan
e-mail: anada-koichi@waseda.jp

T. Kikuchi · T. Yaku
Department of Information Science, Nihon University, Tokyo, Japan
e-mail: kikuchi@yakulab.net

T. Yaku
e-mail: yaku.takeo@nihon-u.ac.jp

S. Koka
College of Humanities and Sciences, Nihon University, Tokyo, Japan
e-mail: koka.shinji@yakulab.net

Y. Miyadera
Department of Mathematics and Information Science,
Tokyo Gakugei University, Tokyo, Japan
e-mail: miyadera@u-gakugei.ac.jp

© Springer International Publishing Switzerland 2016
R. Lee (ed.), *Software Engineering, Artificial Intelligence, Networking and Parallel/Distributed Computing 2015*, Studies in Computational Intelligence 612,
DOI 10.1007/978-3-319-23509-7_10

Koka et al. [2] proposed another method as a modification of the steepest descent line method called the steepest ascent method. This method is based on steepest ascent lines obtained by selecting the maximum inclined direction from eight neighbors of a cell in terrain maps represented by homogeneous rectangular dissections. Then ridge lines are extracted by their steepest ascent lines on a surface.

On the other hands, triangulations of rectangular dissections may be required to effectively display 3D terrain maps. In [3], Kikuchi et. al. introduced a method for triangulation of rectangular dissections and a data structure for generated triangular dissections. This triangulation is suitable to display features of 3D terrain maps such as ridges.

In this paper, we provide a method for ridge detections of terrain maps represented by triangular dissections from the triangulation by [3]. This method is similar to the steepest ascent method by [2] and detects ridge lines along sides of triangles.

In Sect. 2, we survey the steepest ascent method by [2] and a triangulation of rectangular dissections by [3]. In Sect. 3, we introduce an algorithm to detect ridges in triangular dissections on terrain map. In Sect. 4, we compare our methods with one provided in [2] by using an example. Finally, in Sect. 5 we describe conclusion and future works.

2 Related Works

2.1 A. The Steepest Ascent Method for Homogeneous Rectangular Dissections [2]

In [2], the steepest ascent method was proposed to effectively detect ridges. In this paper, we refer to the steepest ascent method in [2] as "*SteepestAscentForRectangle*". Precisely, it is the following algorithm.

ALGORITHM *SteepestAscentForRectangle* [2]

INPUT

- *elevation*: a set of elevation values for all of $m \times n$ cells,
- g_0: a threshold value.

OUTPUT

- *color*: ridge detected map with gray denoted ridge cells.

METHOD

Initialization

- Set $color(i, j) := White$ and $count(i, j) := 0$ for all $i = 1, \ldots, m$ and $j = 1, \ldots, n$.

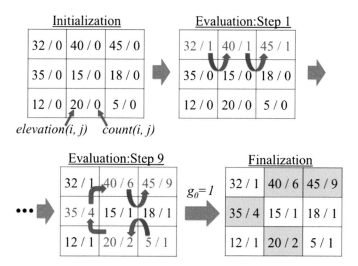

Fig. 1 An example of ridge detection by "*SteepestAscentForRectangle*"

Evaluation

- for all $i = 1, \ldots, m$ and $j = 1, \ldots, n$,
  $count(i, j) + +$ and evaluate $count(i, j)$ by the followings:

  1. Select the (i_{max}, j_{max}) such that (i_{max}, j_{max}) is one of the neighboring cells around (i, j) and satisfies

     $$elevation(i_{max}, j_{max}) = max\{elevation(k, l)|(k, l) \text{ is a neighbor cell.}\}.$$

  2. If $elevation(i_{max}, j_{max}) > elevation(i, j)$
     then
     $count(i_{max}, j_{max}) + +$, replace (i, j) with (i_{max}, j_{max}) and return to line 1.
  3. Else quit.

Finalization

- If $count(i, j) > g_0$ then $color(i, j) := Gray$ for all $i = 1, \ldots, m$ and $j = 1, \ldots, n$.

Figure 1 is an example of ridge detection by the algorithm "*SteepestAscentForRectangle*".

We note that [2, 4] described that the steepest ascent method is better than a method by evaluating the discrete Laplacian.

2.2 A Triangulation of Rectangular Dissections [3]

Kikuchi et al. [3] considered a triangulation to convert rectangular dissections to triangular dissections. In this paper, the method for the triangulation proposed in [3] is referred to as "*TriangulationRectangularDissection*". That is defined as follows:

ALGORITHM *TriangulationRectangularDissection* [3]

INPUT

- D: a rectangular dissection.

OUTPUT

- T: a triangular dissection.

METHOD

- **Step 1**. Put nodes at the center, vertices and the middle of sides of rectangles.
- **Step 2**. Link nodes along sides and connect between the center and boundary nodes.

Figure 2 is an example of triangulations of homogeneous rectangular dissections by the algorithm "*TriangulationRectangularDissection*".

In the algorithm "*TriangulationRectangularDissection*", each of nodes is classified into two types:

"center node" is a node at the center of rectangles

and

"boundary node" is a node on ruled lines is called.

When a homogeneous rectangular dissection has $m \times n$ rectangles, the number of "center node" and "boundary node" is mn and $3mn + 2m + 2n + 1$, respectively. Therefore, the total of nodes generated by "*TriangulationRectangularDissection*" is

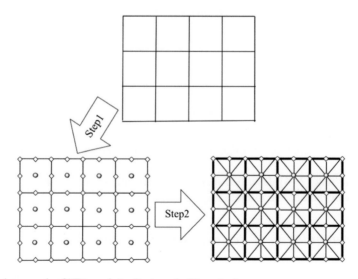

Fig. 2 An example of "*TriangulationRectangularDissection*"

$$4mn + 2m + 2n + 1.$$

This implies that the number of nodes is $O(mn)$ at most.

3 Ridge Detections on Triangular Dissections

Our purpose of this paper is to provide a method for ridge detection on triangular dissections generated by triangulation with "*TriangulationRectangularDissection*".

3.1 Graphs by Triangulation

First, we define undirected graphs to represent triangular dissections generated by applying "*TriangulationRectangularDissection*" to homogeneous rectangular dissections as follows:

DEFINITION

- Let D be a homogeneous rectangular dissection. Then a undirected graph $G_D = (V_D, E_D, h)$ is called a "graph by triangulation for D" if and only if

  - V_D; a set of nodes generated by Step 1 in "*TriangulationRectangularDissection*",
  - E_D; a set links generated by Step 2 in "*TriangulationRectangularDissection*",
  - $h: V_D \rightarrow R$; a real valued function on V_D,

  where the value $h(v)$ has to be defined for any $v \in V_D$. □

  Remark that a real valued function h gives the value of height at each of nodes in V_D to used in ridge detection.

3.2 An Algorithm "SetMeanValue"

In order to generate graphs by triangulation for rectangular dissections, we have to define a real valued function h.

In this paper, we consider rectangular dissections with elevation values. Then, we put elevation values to the values of h on center nodes, that is, if $v \in V_D$ is a center node, then $h(v)$ is equal to the elevation value given at a rectangle corresponding to $v \in V_D$. Next, the value of h on each of boundary nodes is defined as the mean of values at linked center nodes.

Precisely, we provide the following algorithm to generate graphs by triangulation for rectangular dissections with elevation values such as (Fig. 3).

ALGORITHM *SetMeanValue*

INPUT

- D: a rectangular dissection D with $m \times n$ cells.
- *elevation*: elevation values for all of cells in D.

OUTPUT

- G_D: a graph by triangulation for D.

METHOD

- **Step 1**. Apply "*TriangulationRectangularDissection*" to D and then define that

  - $c(i, j)$ is the node at the center of (i, j)-th rectangle,
  - $bh(i, j)$ is the node at the middle of the top side of (i, j)-th rectangle,
  - $bv(i, j)$ is the node at the middle of the left side of (i, j)-th rectangle,
  - $bx(i, j)$ is the node at a intersection (i, j)-th rectangle.

- **Step 2**. Define a set V_D and E_D as

$$V_D = \{c(i, j)|i = 1, \ldots, m \text{ and } j = 1, \ldots, n\}$$
$$\cup \{bh(i, j)|i = 2, \ldots, m \text{ and } j = 2, \ldots, n\}$$
$$\cup \{bv(i, j)|i = 2, \ldots, m \text{ and } j = 2, \ldots, n\}$$
$$\cup \{bx(i, j)|i = 2, \ldots, m \text{ and } j = 2, \ldots, n\}$$

and

$$E_D = \{[v, w] \,|\, [v, w] \text{ is a link between } v \text{ and } w \in V_D$$
$$\text{generated by "\textit{TriangulationRectangularDissection}"}\}.$$

Fig. 3 $c(i, j), bh(i, j),$ $bv(i, j)$ and $bx(i, j)$ on the (i, j)-th rectangle

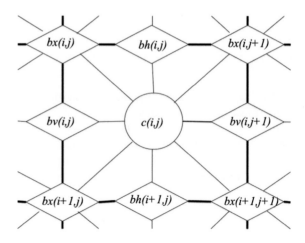

- **Step 3**. Define values of h at $c(i, j)$ as $h(c(i, j)) := elevation(i, j)$ for $i = 1, \ldots, m$ and $j = 1, \ldots, n$.
- **Step 4**. Calculate values of h at $bh(i, j)$, $bv(i, j)$ and $bx(i, j)$ as follows: For $i = 2, \ldots, m$ and $j = 2, \ldots, n$,

$$h(bh(i, j)) := \frac{1}{2}\Big[h(c(i, j)) + h(c(i - 1, j))\Big],$$

$$h(bv(i, j)) := \frac{1}{2}\Big[h(c(i, j)) + h(c(i, j - 1))\Big],$$

$$h(bx(i, j)) := \frac{1}{4}\Big[h(c(i, j)) + h(c(i + 1, j)) + h(c(i, j - 1)) + h(c(i + 1, j - 1))\Big].$$

Figure 4 is an example of a graph generated by "*SetMeanValue*".

3.3 The Steepest Ascent Method Along Sides of Triangules

In this paper, we introduce a method for ridge detections on graphs by triangulation for rectangular dissections. Our method is a modification of "*SteepestAscentForRectangle*" by [2] and finds steepest ascent lines along edges on graphs by triangulation for rectangular dissections.

Precisely, the following algorithm is our method.

ALGORITHM *SteepestAscentForTriangle*

INPUT

- $G_D = (V_D, E_D, h)$: a graph by triangulation for a rectangular dissection D.
- g_0: a threshold value.

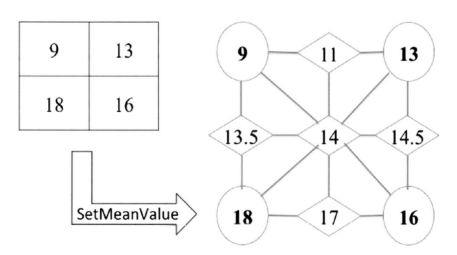

Fig. 4 An example of a graph generated by "*SetMeanValue*"

OUTPUT

- *color*: ridge detected map with gray denoted ridge cells

METHOD

Initialization

- Set $color(v) := White$ and $count(v) := 0$ for all $v \in V_D$.

Evaluation

- For all $v \in V_D$,
  $count(v) + +$ and evaluate $count(v)$ by the followings:

  1. Select $v_{max} \in V_D$ such that $[v_{max}, v] \in E_D$ and satisfies

  $$h(v_{max}) = max\{h(w)|w \in V_d \text{ and } [w, v] \in E_D\}.$$

  2. If $h(v_{max}) > h(v)$, then
     $count(v_{max}) + +$, replace v with v_{max} and return to line 1.
  3. Else quit.

Finalization

- For all $v \in V_D$, if $count(v) > g_0$ then $color(v) := Gray$.

  Figures 5 and 6 are examples of ridge detections by "*SteepestAscentForTriangle*".

Remark When a map is represented by homogeneous rectangular dissections, we first apply "*SetMeanValue*" and then ridge lines are detected by "*SteepestAscentFor-Triangle*".

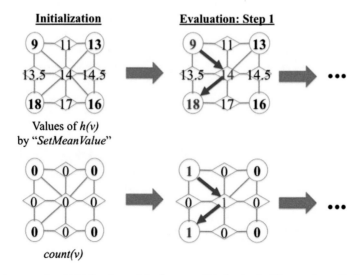

Fig. 5 An example of Initialization and the first step of Evaluation in "*SteepestAscentForTriangle*"

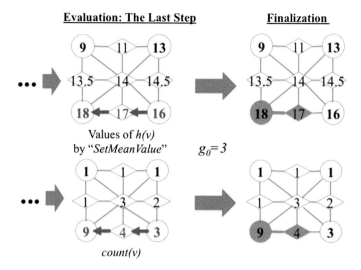

Fig. 6 An example of Finalization in "*SteepestAscentForTriangle*"

In this section, we provided two algorithms. The first is "*SetMeanValue*" to cal-culate values of h at boundary nodes and the second algorithm "*SteepestAscent-ForTriangle*" is to detect ridge lines in maps generated by "*SetMeanValue*". This implies that we can independently develop algorithms for calculations of values of h at baoundary nodes and detections of ridges and other features in maps.

4 Comparison

We compare our method for graphs by triangulation with a result by algorithm in [2] in an example. That is, a homogeneous rectangular dissection with elevation values is given as an example and we apply "*SetMeanValue*" and "*SteepestAscentForTriangle*" provided in the previous section. Then we will compare the results with one by "*SteepestAscentForRectangle*".

In this section, let D be a homogeneous rectangular dissection with the elevation values in Fig. 7.

First, we apply our method to this example. First, a rectangular dissection D is converted to a graph by triangulation for D, that is, we apply the algorithm "*Set-MeanValue*" to D. Then we can get a graph in Fig. 8.

And then we apply the algorithm "*SteepestAscentForTriangle*" to a graph given in Fig. 8.

The result is Fig. 9. Note that detected ridge lines are clear connected between gray nodes. Precisely, the result by our method has two ridge lines in this example.

32	40	45
35	15	18
12	20	5

Fig. 7 An example of a rectangular dissection D and values of *elevation*

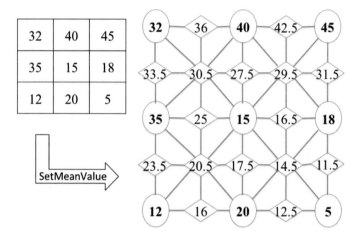

Fig. 8 A graph generated by applying "*SetMeanValue*"

Next, we apply "*SteepestAscentForRectangle*" by [2] to D. Then we can get Fig. 10. Note that the threshold value in "*SteepestAscentForRectangle*" is $g_0 = 1$.

Detected ridges in Fig. 10 may not have clear lines even thought our method gives clear ridge lines. In Fig. 11, Fig. 10 is fitted to Fig. 9 overlaped with corresponding cells.

In this example, we show that our method can give clearer ridge lines than ones by "*SteepestAscentForRectangle*" introduced in [2].

5 Conclusion

In this paper, we define graphs by triangulation for rectangular dissections. And then we provide an algorithm to generate graphs by triangulation for rectangular dissections and a method to detect ridge lines. Our method is a modificationof the

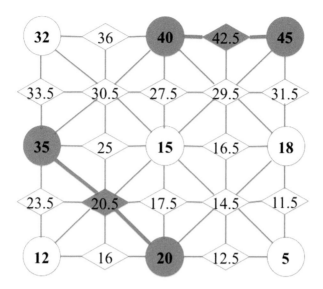

Fig. 9 The result by applying "*SteepestAscentForTriangle*"

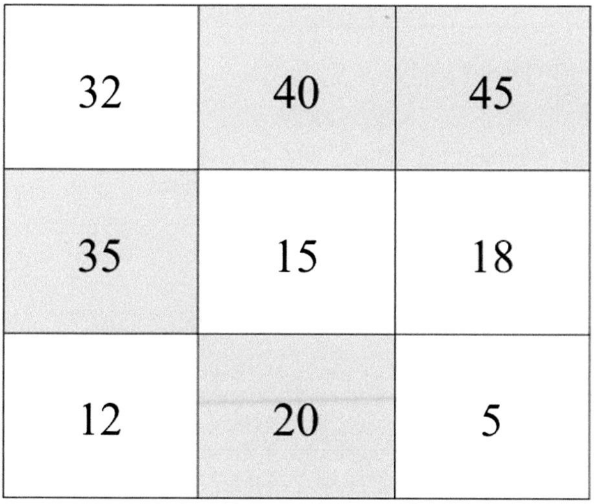

32	40	45
35	15	18
12	20	5

Fig. 10 The result by applying "*SteepestAscentForRectangle*"

steepest ascent method for homogeneous rectangular dissections given in [2] and can give ridge lines in terrain map represented by homogeneous triangular dissections.

As future works, we improve our method to sharpen features such as ridges and apply this method to triangular dissections generated by heterogeneous rectangular dissections.

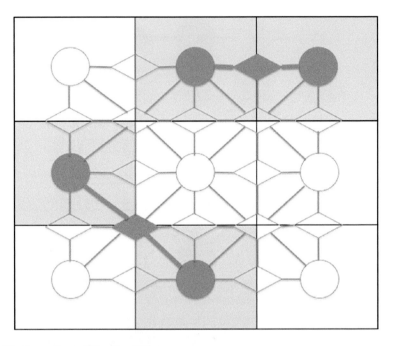

Fig. 11 Comparison of Figs. 9 and 10

Acknowledgments We would like to thank Prof. Kimio Sugita of Tokai University and Prof. Kensei Tsuchida of Toyo University for their valuable suggestions.

References

1. Cousty, J., Bertrand, G., Najman, L., Couprie, M.: Watershed cuts: minimum spanning forests and the drop of water principle. Proc. IEEE Trans. PAMI **31**(8), 1362–1374 (2009)
2. Koka, S., Anada, K., Nomaki, K., Sugita, K., Tsuchida, K., Yaku, T.: Ridge detection with the steepest ascent method. procedia Comput. Sci. **4**, 216–221 (2011)
3. Kikuchi, T., Anada, K., Koka, S., Miyadera, Y., Yaku, T.: A data structure for triangular dissection of multi-resolution images. In: Proceedings of the 15th ACIS International Conference on Software Engineering, Artificial Intelligence, Networking and Parallel/Distributed Computing (SNDP 2014), pp. 490–495 (2014)
4. Koka, S., Anada, K., Nakayama, Y., Sugita, K., Yaku, T., Yokoyama, R.: A comparison of ridge detection methods for DEM data. In: Proceedings of the 13th ACIS International Conference on Software Engineering, Artificial Intelligence, Networking and Parallel/Distributed Computing (SNDP 2012), pp. 513–517 (2012)
5. Yokoyama, R., Kureha, A., Motohashi, T., Ogasawara, H., Yaku, T., Yoshino, D.: Geographical concept recognition with the octgrid method for learning geography and geology. Proc. IEEE ICALT **2007**, 470–471 (2007)

Architecture for Wide Area Appliance Management

Arata Koike and Ryota Ishibashi

Abstract We studied architecture for Internet-of-Things (IoT) appliances with constrained resources to enable controlling and managing them over a wide area network. By clarifying requirements for using wide area network, we examined issues that are associated with each related standard-based technologies. Our analysis gives us a solution by combining CoAP and ECHONET Lite to complement each other to overcome the issues associated with using them in a wide area network, especially suitable for virtual gateways located in a cloud. We then showed a realization of our proposed architecture by prototyping the system.

Keywords CoAP · ECHONET Lite · Internet of things · Wide area appliance management

1 Introduction

Lightweight and compact IP (Internet Protocol) technologies have been carried out to support devices with constraint processing and communication capabilities [1–4]. In addition to the HTTP [5] etc., which are used in conventional Internet applications, we have new application layer protocols such as CoAP [1] and MQTT [6]. CoAP is especially tuned up for constraint IoT/M2M devices with lightweight implementation. MQTT also has a lightweight footprint aimed for message queue type architecture. These protocols are suitable for lightweight device management that works with large-scale, wide area environments, such as on the Internet. However, they are designed for general purpose, so we do not have a specific data model for actual

A. Koike (✉) · R. Ishibashi
Network Technology Laboratories, Nippon Telegraph and Telephone Corp.,
Tokyo 180-8585, Japan
e-mail: koike.arata@lab.ntt.co.jp

R. Ishibashi
Currently with Service Design Division, NTT DOCOMO, Inc.,
Tokyo 100-150, Japan
e-mail: ryouta.ishibashi.rc@nttdocomo.com

© Springer International Publishing Switzerland 2016   143
R. Lee (ed.), *Software Engineering, Artificial Intelligence, Networking
and Parallel/Distributed Computing 2015*, Studies in Computational Intelligence 612,
DOI 10.1007/978-3-319-23509-7_11

device control with them. Therefore, we have to newly define data model or to import existing ones.

Home appliances are typical examples of the constraint devices on processing and communication capabilities. In Japan, ECHONET Lite [7] is the most widely recognized standard protocol for controlling and managing these home appliances. ECHONET Lite is also accepted as a standard for smart home and smart energy management and we expect the growth of appliances implementing it. ECHONET Lite has detailed data model to control ECHONET enabled appliances. It, however, focuses on a local area network such as a home network so it is not easy to utilize it for a large-scale wide area network. Therefore, the ECHONET Lite is currently used within a local area network where both home appliances and a dedicated management system coexist.

If we can utilize ECHONET Lite over wide area network such as telecom network or the Internet, we do not need to install a dedicated management system in a local network environment. Instead, we can place a management system on a cloud data center and can provide a cloud-based appliance management service. The cloud-based service makes it easy for simplifying configuration of a home network. It can also provide flexible service update and collaboration with other services on the cloud [8]. This gives great benefit for both users and service providers. If we consider virtual Customer Premises Equipment (vCPE) or a virtual home gateway in a cloud, this simplifies boxes in a home. To realize this, we have to find a way on how to manage appliances without having any gateways or middle boxes with richer capabilities in a home environment.

This paper proposes architecture combining ECHONET Lite and CoAP to enable a large-scale, wide area cloud based home appliance management service. This proposed architecture complements ECHONET Lite and CoAP each other and utilizes their strength. ECHONET Lite and CoAP have different features but they are not exclusive ones. If we can carefully combine both of them, the proposed architecture has a potential to enhance their strength. The main outcome of this proposal is to extend the ECHONET Lite system for resource constraint appliances, which currently works within a local area network, to a cloud-based wide area system. As we have to take into account the resource constraint for the home appliances, CoAP plays an essential role to keep the low implementation footprint.

This paper is organized as follows. In Sect. 2, we overview researches for managing home appliances over wide area network. And then we summarize features for CoAP protocol, which was developed for constraint devices, and ECHONET Lite, which is designed for management and control for home appliances. In Sect. 3, we propose a novel architecture for controlling home appliances in a home network through a wide area network by ECHONET Lite over CoAP. Section 4 describes our prototyping work to demonstrate the proposed architecture and we show that our proposed architecture enables end-to-end communication for appliance management. We discuss our observations in Sect. 5. Section 6 concludes the paper.

2 Managing Home Appliances over Wide Area Network

There are a number of studies conducted on controlling M2M/IoT devices through a wide area network [8, 9]. We will first review several previous works. Then, also review CoAP and ECHONET Lite with their features. We then describe technical challenges associated with the use of ECHONET Lite through a wide area network.

2.1 Previous Works for Controlling Home Appliances Through a Wide Area Network

There are a number of approaches for controlling home appliances through wide area network [10]. Several standards exist for them. In [11], it discusses so called Multi-Prefix Multi-Homing method, where different IPv6 prefixes are distributed for home appliances depending on the service providers. Home Gateway Initiative [12] specifies their architecture to place application gateway in a home gateway based on OSGi [13] framework. In this case, a home gateway terminates home networking protocols and initiates a different protocol to communicate with applications on the cloud. Similar approaches are widely used to establish an interworking function at application layer to correlate a closed local area network and an open wide area networks [14]. In the methods above, we need an application in a gateway box to establish end-to-end control and management capabilities.

In [15], they encapsulate ECHONET in HTTP for end-to-end as a way to enable wide area Home Energy Management System (HEMS). They discuss their approach based on the assumption that M2M/IoT appliances are capable with large frame processing. If we consider resource constraint devices, however, we have to consider not only frame length but also various overheads such as protocol processing as described in [3]. When we use HTTP, we have to consider not only the frame length and processing for HTTP itself, but also consider the lower layer TCP/IP protocol processing. This actually affects implementation footprint. In [3], home appliances are categorized Class 1, where we assume resources with less than 10 kbyte RAM and less than 100 kbyte program code. HTTP on top of TCP/IP apparently exceeds this class.

2.2 Communications by Devices with Resource Constraint

The Internet Engineering Task Force (IETF) recently studies IP-based communications for devices with resource constraint as in [3] actively for several layers. For application layer, core Working Group specifies CoAP protocol [1] as an application protocol for M2M/IoT environment based on Representational State Transfer (REST) architecture.

CoAP is designed to work on severe resource constraint environment with the specific requirements. Following are its major requirements:

- Lightweight footprint to implement on scarce resource environment
- Support of sleep node
- Low latency and small processing requirement in constraint communication environment and capable of reliable communication
- Mapping to HTTP and vice versa

CoAP architecture consists of three types of entities: client, server, and proxy. REST architecture is used for designing CoAP protocol. It has Create, Retrieve, Update, and Delete (CRUD) operations for resources designated by URIs. Similar to HTTP, Proxy is also assumed. A Proxy can relay CoAP message and also can provide interworking with HTTP. This HTTP friendliness is one of the features for CoAP in both architecture and protocol aspects. CoAP is basically designed for low power protocols such as [2] but it does not depend on any payload, Layer 3 nor lower protocols except it utilizes UDP (/DTLS) for its transport protocol.

CoAP has internally two virtual layers structure. It can provide its own transport function and REST capabilities by splitting transaction for CoAP message and request/response for REST layer.

- REST layer: it provides REST function by methods such as GET/PUT to operate resources designated by URIs.
- Transaction layer: it is located below REST layer and provides connectivity management on the UDP. It corresponds to SYN, and ACK for TCP.

Shelby et al. [1] defines a proxy server function. Similar to HTTP, the proxy server function relays and caches CoAP messages. It also defines cross protocol proxy, where it converts mutually between HTTP and CoAP messages.

Unlike HTTP or SIP, which is text based message format, CoAP is a binary coded protocol. It has fixed 4 byte header and there is no mandatory options for CoAP. Therefore, the minimum message size of CoAP is 4 byte when it comprises only a header and no body. It uses UDP for its transport protocol but has own control mechanisms for flow and reliability. The explanation below is a brief comparison of CoAP control mechanism with that of TCP. Since CoAP does not assume transactions of large size data so it does not have a window-based control mechanism, by which TCP relies on. CoAP has different philosophy for congestion control than TCP. TCP relies on slow-start mechanism to control the number of packets in a connection. On the other hand, CoAP avoids increase of flowing packets by controlling the number of simultaneous connections to a single server. It also has a mechanism to randomize responding timings for multicast to avoid simultaneous reply transmission. For the CoAP retransmission mechanism, it has a maximum number for retransmission try and exponential back-off mechanism to double the retransmission timer when it retransmits. By combining them, it controls retransmission of packets. It chooses a random value from a certain range for the initial value of the retransmission timer and thus it avoids synchronization of retransmissions among multiple clients. Unlike TCP, CoAP does not have upper limit for timer but it stops retransmission when number of retransmission reaches the maximum value (the default value is 4.) TCP changes its timer value during communications by measuring Round Trip Times (RTT) but CoAP does not.

In order to look at the lightness of CoAP, we compare data size between HTTP and CoAP using an assumed typical communication model described below. We model a round trip sequence of a creation (POST) of a resource and a reply (201) message of it with standard header and options. For HTTP with standard header structure, request is 247 bytes and response is 162 bytes. For CoAP, they are 23 bytes and 4 bytes, respectively. This is 2.5 $\sim$ 10 % of the size for HTTP. If we include all overheads below TCP or UDP, CoAP packet size is approximately 25 % of HTTP. CoAP uses binary format to achieve effectiveness of header space. This effectiveness will decrease by the increase of the portion of URI and payload, which does not obtain gains by binary coding. We calculate payload size, message size and ratio of them for standard HTTP and CoAP request including header option. When payload size is less than 32 bytes, CoAP can keep its message size less than 20 % of HTTP. If payload size exceeds 256 bytes, CoAP/HTTP ratio exceeds 50 % and it loses size reduction effect. CoAP uses UDP as its transport layer. So it does not need a three-way handshake as in HTTP/TCP. If we assume no retransmission, one round-trip of request and response produces 11 packets for HTTP/TCP but 2 packets for CoAP/UDP. This is 1/5 of HTTP/TCP and improves efficiency. Note that TCP usually sends several segments in a connection, so in this case, TCP improves efficiency. Those are features of CoAP and we can see that it maintains compatibility with HTTP but achieve lightness to support M2M/IoT with resource constraints. So this compatibility gives us anticipation to apply CoAP to wide area network.

2.3 ECHONET Lite and Issues for Applying it to Wide Area Network

ECHONET Lite is an application layer protocol. It defines objects by abstracting appliances such as light or air conditioner. Then it defines precise data structure that represents settings and status of each object. This enables retrieval or control of values or status of the object from a controller. ECHONET Lite enables remote monitoring of appliances by interpreting signals based on the defined data structure. This data structure describes ECHONET Lite frame. It can be conveyed over any transport layer protocol including TCP/UDP (Fig. 1 left-side). Usually, typical home appliances are not tuned as communication equipment. So they do not have enough assigned resources for control, manage, and status check. This lack of resources leads simplified communication procedures without retransmission, authentication, and security protection. And it reduces power consumption and equipment cost. As a result, ECHONET Lite lacks following key functions to use it in a wide area network, such as the Internet.

Fig. 1 Comparison of
protocol stacks between
ECHONET Lite and
proposed scheme

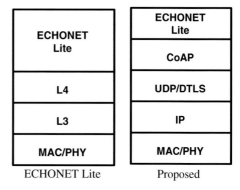

ECHONET Lite   Proposed

(A) Retransmission control

Retransmission is left for transport layer protocol or application ones. This
will not cause a big problem if we use this for a small home area network
environment, where we can expect small packet loss ratio. Even if we suffer
packet loss events and the communication is not completed, appliance managers
or management applications can detect them in a limited local area network,
where only a limited number and kinds of appliances exist. And they can easily
take appropriate actions such as resubmission of the job since they could see
what was the result of their action. When we use it through a wide area network,
the network size becomes larger and packet loss possibility will increase. And
we could not see or identify the result of actions in this case. So it is better to
have retransmission capability within the protocol.

(B) Address resolution

ECHONET Lite does not provide a scheme to centrally manage the addresses,
types, and capabilities of target appliances in the network. When we use UDP
as its transport layer, each appliance uses IP multicast to advertise its address,
type, and capability information. Other appliances can know this information by receiving this IP multicast. Unfortunately, only limited Internet service
providers (ISPs) allow the use of IP multicast in their network and it is used
only limited purposes, such as video distribution for IPTV within an ISP. We
have to say it is difficult to use IP multicast across several ISPs right now. This
makes it difficult to use ECHONET through a wide area network. So we need
different mechanisms for address resolution if we use it through a wide area
network.

(C) Resource Discovery

At the same time with address resolution, a node advertises and learns types and
capabilities using IP multicast. In wide area networks, we need to investigate
how we can make this resource discovery.

As described above, Since ECHONOET Lite is targeting local network environment, there are problems A to C above for applying it for a wide area network.

3 ECHONET Lite over CoAP for Wide Area M2M/IoT Appliance Management

As we look at in Sect. 2, both CoAP and EHONET Lite are lightweight protocols for M2M/IoT appliances with scarce resources. Both of them have problems if we use it through a wide area network. In this section, we propose an architecture to combine ECHONET Lite and CoAP to overcome these problems associated with using it in a wide area network. We first analyze and compare both protocols from the viewpoints of architecture, data access, protocol, Identifier, and data model.

3.1 Comparison in Architecture

CoAP is client-server type architecture. Client sends requests and server processes receiving requests and both are completely separated. ECHONET Lite is actually comprised of a controller where sends a controlling request, and devices where receives and processes requests and are controlled by the requests. However, in the specification, each ECHONET Lite node is an equal entity so we can say it is a peer-to-peer type. CoAP is designed to accommodate proxy servers, like HTTP, to relay messages. On the other hands, ECHONET Lite is not designed to support proxy functions. CoAP must have both client and server function and need to use them properly. While in ECHONET Lite, both kind of functions works simultaneously. Therefore, if we want to have both autonomous data read and write functions and a function accepting an incoming control message from outside, ECHONET Lite is relatively easy to implement.

3.2 Comparison of Data Access

CoAP adopts REST style data access, i.e., it specifies server or resource using an URI and manipulates it by a method such as GET, PUT. ECHONET Lite seeks object-oriented style. It models appliances as 'Objects.' An Object has 'Property'. When needed, we identify Object and Property by ID, and manipulate (such as read, write) the Property. CoAP identifies the location (IP address) of the resource using URI and DNS. On the other hand, ECHONET Lite does not have any mechanisms to identify an address of an appliance. This means CoAP needs only existing schemes (i.e., IP, URI, and DNS) for web but ECHONET Lite needs to rely on other method to identify a peer to communicate.

3.3 Comparison of Identifier

CoAP identifies a manipulating resource using an URI. An URI is comprised of a combination of FQDN (IP address) that designates a host and an URI path that designates a resource. An URI gives a globally unique resource location but a resource could have multiple URIs (a resource having this multiple URIs affects cache behavior.) ECHONET Lite uses 'Object' to represent appliance types (such as air conditioner or television) and 'Property' to represent function of appliances (such as ON and OFF). This abstraction gives the possibility to have the same Object IDs to the same type of appliances and the same Property ID for the same or similar functions among appliances. Therefore, we cannot identify a target appliance by Object ID or Property ID only.

3.4 Comparison of Protocol

CoAP uses binary format and have a fixed 4 byte mandatory header and variable length options using Type-Length-Value (TLV) encoding, and a payload. It uses UDP for its transport layer and has its own retransmission mechanism. ECHONET Lite also uses a binary format and a fixed mandatory 4 bytes header. It can have either combination of predefined sub header and payload or combination of arbitrary format and length of payload. There is no option field. ECHONET Lite does not assume specific transport layer but many implementations use UDP. There is no specification of retransmission behavior on UDP for ECHONET Lite. CoAP has more flexibility by using option than ECHONET Lite and has potential to achieve stability among nodes using retransmission mechanism when used in a wide area network.

3.5 Comparison of Data Model

CoAP does not specify any specific data model to express status of resources and it depends on application. It provides a mechanism (CoRE Link Format) to refer resource types or interfaces but it does not specify an interface itself. It does not concern how actually an appliance runs as a result of resource operation.

For ECHONET Lite, it rigidly defines appliance (class) and functions of the appliance (property) as a data model. Therefore, it strictly defines what kind of appliances can be controlled by which method.

We believe that strict ruling approach like ECHONET lite contributes enhancing interoperability among appliances especially in early market stage. It, on the other hand, has a drawback that we have to define class and property for any appliances so that it creates some delay to support new appliance or functions. For CoAP, how to support each appliance is completely application specific. So it increase risks to create numerous similar but different data models and this leads to serious interoperability issues unless we find a de facto standard. Note that CoAP can co-exist with strict data model approach, as it does not specify data model itself.

3.6 Effectiveness of Protocol Integration

To summarize the observations above, we can say that CoAP is appropriate to use with wide area networks, it has flexibility to support new appliances or new functions. However, it has a risk on interoperability. For ECHONET Lite, it focuses on local network and it specifies strict meaning of data for appliances and functions. So it has better interoperability with the sacrifice of delay in supporting new appliances or functions. These consideration leads to a solution that ECHONET Lite over CoAP can maintain benefits of the both protocols and mitigate demerits of them, especially when we use it through a wide area network. More accurately, we treat CoAP as a transport and ECHONET Lite as a data model. This approach compensates drawbacks of each protocol. In Fig. 2, we show a concept of cloud based appliance management using ECHONET Lite over CoAP and illustrate major benefits.

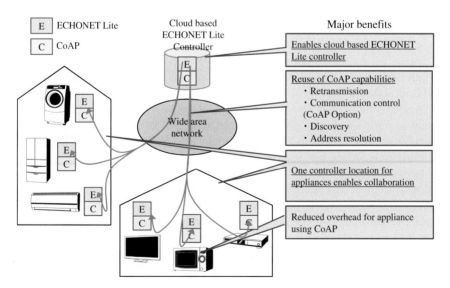

Fig. 2 Concept of cloud-based system using ECHONET Lite over CoAP

Fig. 3 The proposed architecture

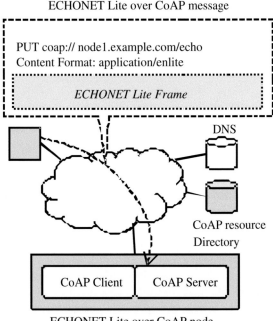

ECHONET Lite over CoAP message

PUT coap:// node1.example.com/echo
Content Format: application/enlite

ECHONET Lite Frame

DNS

CoAP resource
Directory

CoAP Client   CoAP Server

ECHONET Lite over CoAP node

3.7 Proposed Architecture

We show protocol stack and architecture of our proposed ECHONET Lite over CoAP in Fig. 1, right and Fig. 3.

A node capable of the proposed architecture sends and receives ECHONET Lite frame as a payload for a CoAP message. A node acts as a client when it sends or receives any frames with ECHONET Lite service (ESV) other than responds such as read, write, notify, etc. It creates and sends a CoAP request with that frame as its payload. When a node receives the CoAP message, it acts as a CoAP server and extracts the frame and passes it to ECHONET Lite layer. The ECHONET Lite layer processes that message in the frame and the result will become ECHONET Lite response. The corresponding CoAP response message is created and ECHONET Lite response is stored in the payload of it and returned to the originating client. To identify that the payload has ECHONET Lite frame in a CoAP message, we set Media-Type (e.g., *application/enlite*) in the Content-Format option field.

A CoAP server that receives requests in a node can have multiple resources specified by URLs. In our proposed architecture, there are two cases: one resource processes all EHONET Lite appliances' objects in a node, and multiple resources process the corresponding appliance's object. For the former case, we use [16] and define a common URL path (e.g., */.well-known/echonet-lite*) as a resource to process CoAP message. In the latter case, we adopt [17] and define the *rt* attribute that identifies resource types and express the support of the proposed architecture for each resource (e.g., $rt = enlite$). Each node registers and manages the above resource information that it hosts in accordance with the procedure in [18]. A node that directs how to control and manage discovers address and resource information of the other nodes that process the CoAP message for control and manage by using [19]. When address resolution is needed during message transmission, CoAP layer utilizes existing mechanisms such as DNS.

By this proposed architecture, we can meet required functions A, B, and C described in Sect. 2.3 using functions in CoAP or lower layer. This means we can eliminate burden from ECHONET Lite layer or an application. We then can utilize ECHONET Lite in IP reachable network environment including wide area network. Therefore, we can achieve appliance management through a wide area network using ECHONET Lite.

4 Consideration by Prototyping of ECHONET Lite over CoAP

We consider the realization of our proposed architecture in Sect. 3.7 by using a prototyping implementation. First, we studied investigation of problems associated with ECHONET Lite over CoAP implementation. Then we consider how we can map functions between both protocols.

Our aim of prototyping is not for direct commercial usage but trying to establish a reference model for future deployment. For this purpose, we utilize open source software for our basis. And our target is home appliances so we choose C/C++, which is lighter than Java, for our development language.

There are several different CoAP implementations. We use libcoap [19] as this has no strict bonding by licensing policy and based on the most up-to-date CoAP specifications. As for ECHONET Lite, we use OpenEcho [20] as this is the only available open source implementation. Note that [20] is Java-based implementation so we convert it to C++ by using Java to C++ converter [21].

We developed a mapping method of ECHONET Lite objects to CoAP as shown in the rule in Table 1 for our implementation. Using our implementation, we tested our architecture based on the sequence in Fig. 4. We confirmed that our proposed architecture works based on the sequence.

5 Discussion

For actual device control such as turning lights on and off, we need not only communication layer processing software but also application software, by which we can achieve logical resource control and physical hardware control. CoAP

Table 1 Mapping of ECHONET Lite services

Service Code (ESV)	ECHONET Lite service	symbol	remark	mapping
0x60	Property value write request (no response required)	SetI		POST
0x61	Property value write request (response required)	SetC		POST
0x62	Property value read request	Get		GET
0x63	Property value notification request	INF_REQ		GET
0x64–0x6D	for future reserved			
0x6E	Property value write & read request	SetGet		GET
0x6F	for future reserved			
0x71	Property value Property value write response	Set_Res	Response to ESV=0x61	POST
0x72	Property value read response	Get_Res	Response to ESV=0x62	POST
0x73	Property value notification	INF	Spontaneous property value notification or use for response to 0x63	POST
0x74	Property value notification (response required)	INFC		POST
0x75–0x79	for future reserved			
0x7A	Property value notification response	INFC_Res	Response to ESV=0x74	POST
0x7B–0x7D	for future reserved			
0x7E	Property value write & read response	SetGet_Res	Response to ESV=0x6E	POST
0x7F	for future reserved			
0x50	Property value write request "response not possible"	SetI_SNA	Invalid response to ESV=0x60	POST
0x51	Property value write request "response not possible"	SetC_SNA	Invalid response to ESV=0x61	POST
0x52	Property value read "response not possible"	Get_SNA	Invalid response to ESV=0x62	POST
0x53	Property value notification "response not possible"	INF_SNA	Invalid response to ESV=0x63	POST
0x54–0x5D	for future reserved			
0x5E	Property value write & read "response not possible"	SetGetI_SNA	Invalid response to ESV=0x6E	POST
0x5F	for future reserved			

protocol is a general purpose protocol so it does not specify application part. However, there are various kinds of M2M/IoT devices, so not the communication part but the implementation of device application could be the most important factor for the whole developing costs. In this context, it is beneficial for developers, who are familiar with web developing, to give an http-like environment by spinning off the communication parts from application. CoAP provides these communication parts. Then the developers only need to focus on application parts and they do not need to make their own 'application' including communication part. ECHONET Lite is an international standard for home appliance managements. It provides logical control method for home appliances. In recent years, many appliances begin to support ECHONET Lite. So if developer can utilize both standard based technologies, we can say that it has a potential to extend the developer community and promote the wide adoption of CoAP and ECHONET Lite devices.

In our prototyping experience described in Sect. 4, we confirmed that we can easily implement general resource operation by CoAP as we do in HTTP. We also confirmed that CoAP has compatibility with HTTP. This means it reduces bars for M2M/IoT development for Web developers. This is one benefit of our ECHONET over CoAP architecture.

Other benefit is that this gives a solution for realizing virtual CPE (vCPE). Virtual CPE is a concept to locate CPE or a home gateway not in a home but in a cloud. This means a home network domain is logically extended to a cloud through a wide area network. Number of networked appliances in a home increases in near future, and those need controllers. If we want to use local boxes for these controllers, we need

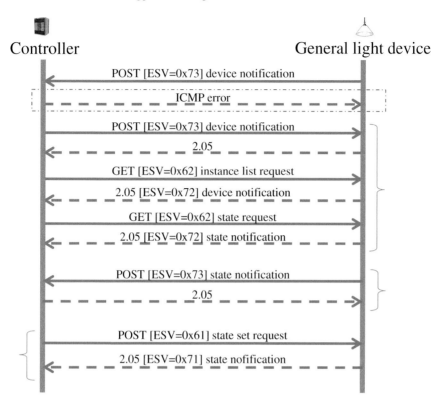

Fig. 4 An example sequence for testing

multiple controller boxes or a very high performance box to install various controller applications in a home. To simplify this situation, vCPE gives a solution. When we use vCPE in a cloud, we have to manage appliances through a wide area network. When we communicate through a wide area network, we have requirements A–C given in Sect. 2.3. Our proposed ECHONET Lite over CoAP architecture is designed to meet these requirements so we can say our architecture is suitable for supporting vCPE with remote appliance management capabilities.

6 Conclusion

In this paper, we studied the architecture to control and manage appliances with resource constraint through a wide area network, such as the Internet. We first clarify the requirements for remote control through a wide area network. Then, we analyze the standardized protocols and identify their difficulties when applying them separately for remote control. We then propose to utilize CoAP as a transport function and ECHONET Lite as a data model function. We show that this combination can solve difficulties associated with controlling appliances through a wide area network. And we thus establish an architecture to control resource constraint home appliances through a wide area network by standardized protocol with our prototyping.

There are ongoing discussions for M2M platform at standardization groups such as OneM2M [22]. We will consider remote appliance control for larger scale environment by relying on the discovery functions provided by the M2M platform.

Acknowledgments We thank Noi Koike and Rumina Koike for proof reading this manuscript.

References

1. Shelby, Z., Hartke, K., Bormann, C.: The Constrained Application Protocol (CoAP). RFC7252, IETF (2014)
2. Hui, J. (ed.), Thubert, P.: Compression Format for IPv6 Datagrams over IEEE 802.15.4-Based Networks. RFC6282, IETF (2011)
3. Bormann, C., Ersue, M., Keranen, A.: Terminology for Constrained-Node Networks RFC7228, IETF (2014)
4. OMA: Lightweight Machine to Machine Technical Specification. OMA-TS-LightweightM2M-V1_0-20131210-C, Open Mobile Alliance (2013)
5. Fielding, R., Gettys, J., Mogul, J., Frystyk, H., Masinter, L., Leach, P., Berners-Lee, T.: Hypertext Transfer Protocol–HTTP/1.1. RFC2616, IETF (1999)
6. International Business Machines (IBM) Corporation, Eurotech: MQ Telemetry Transport (MQTT) V3.1 Protocol Specification (2010)
7. ECHONET Consortium: ECHONET Lite Specification, version 1.1 (2014)
8. Toji, R.: Trends concerning standardization of openADR. NTT Tech. Rev. 11(12) (2013)
9. Kitano, R., Tatemichi, H., Iwasaki, N., Toji, R.: Automated appliance control with cooperation of automated demand response and conditions. In IEICE Communication Society Conference, 210, B-9-11, September (2013)
10. Fujita, T., Goto, Y., Koike, A: M2M Architecture Trends and Technical Issues (in Japanese). Journal of IEICE Vol.96, No.5, pp.305-312, May (2013)
11. IPv6 Promotion Council: Consideration on how to construct IPv6 multi-prefix environment (in Japanese) (2007)
12. Home Gateway Initiative: Use Cases and Architecture for a Home Energy Management Service. HGI-GD017-R3 (2011)
13. OSGi Alliance Web Page. http://www.osgi.org/Main/HomePage
14. Akai, K., Fukuda, F., Fukushima, N., Yanata, R., Furukawa, Y.: A study on device management method using IEEE1888 in smart communities (in Japanese). IPSJ, SIG Tech. Rep. vol. 2012 CDS-4(10) (2012)
15. Masuo, T., Nakamura, J., Matsuoka, M., Hasegawa, G., Murata, M., Matsuda, K.: Study on HEMS over cloud system utilizing realtime web technologie (in Japanese). IEICE Tech. Rep. (NS2012-117) **112**(350), 1–6 (2012)
16. Nottingh, M., Hammer-Lahav, E.: Defining Well-Known Uniform Resource Identifiers (URIs). RFC5785, IETF (2010)
17. Shelby, Z.: Constrained RESTful Environments (CoRE) Link Format, RFC6690, IETF (2012)
18. Shelby, Z., Bormann, C., Krco, S.: CoRE Resource Directory. Internet-Draft, draft-ietf-core-resource-directory-05, IETF (2015)
19. libcoap: C-Implementation of CoAP. http://sourceforge.net/projects/libcoap/
20. Sony Computer Science Laboratories, Inc.: OpenECHO. https://github.com/SonyCSL/OpenECHO

21. Java to C++ converter. https://code.google.com/a/eclipselabs.org/p/j2c/
22. OneM2M. http://www.onem2m.org

Towards a Model Level Replication Technique for Fault Tolerant Systems Using AADL

Wafa Gabsi and Bechir Zalila

Abstract The replication, a technique widely used for fault tolerance purposes, is defined as the redundancy of software, hardware or both units and their consideration in the execution of the application. In this paper, we propose a new technique to design replication using the AADL language and its extensibility with property sets. We choose AADL to take advantage of its strong semantics at architecture level. We enable the designer to model his application using AADL and to enrich it with the property set `Replication_Properties`. We defined this property set to describe the adopted concepts of replication. Then, based on a set of transformation rules, we generate an intermediate AADL model enriched with different replicas. Currently, we are extending the Ocarina tool suite to support automatic generation of the target model.

1 Introduction

As the real-time critical systems are more and more complex and evolved, new requirements for high dependability, fault tolerance and error recovery emerge. To cope with this evolution, researchers focus on how to guarantee the dependability of such systems since the design level.

The dependability is defined as the ability to deliver a service that can be justifiably trusted. There are several means to ensure dependability such as fault tolerance which is defined as the capability of a system to continue providing services even in the presence of errors [1].

A widely used technique to achieve fault tolerance is the replication [10]. This technique involves repetition and multiplicity of different or symmetric components

W. Gabsi (✉) · B. Zalila
ReDCAD Laboratory, National School of Engineers of Sfax, University of Sfax,
B.P. 1173, 3038 Sfax, Tunisia
e-mail: wafa.gabsi@redcad.org

B. Zalila
e-mail: bechir.zalila@enis.rnu.tn

© Springer International Publishing Switzerland 2016
R. Lee (ed.), *Software Engineering, Artificial Intelligence, Networking and Parallel/Distributed Computing 2015*, Studies in Computational Intelligence 612,
DOI 10.1007/978-3-319-23509-7_12

or treatments. It consists on considering multiple copies of a software or hardware components and deploying them on different nodes in order to avoid the failure of the system. Thus, critical hardware or software components, or even entire systems, are replicated. Three main replication techniques are distinguished in the literature [10]:

1. *Active replication*: All replicas have the same inputs, keeping their internal state synchronized and voting all on the same outputs. In this case, we must have a voting algorithm to choose one between all the outputs.
2. *Passive or primary-backup replication*: Only one replica, called primary copy, can ensure the inputs treatment. When the primary copy fails, one of the others, called backup copies, is elected to take its place to provide the same functionality.
3. *Semi-Active replication*: Similar to the active one, all replicas receive the same inputs and can thus treat them. However, similar to the passive replication, there is a privileged copy (the primary) responsible for taking decisions without needing a consensus algorithm to vote between replicas.

There are several work aiming at modeling or implementing fault tolerance techniques based on replication. In our context, we deal with active and passive replication of both hardware and software components. We have chosen AADL (Architecture Analysis & Design Language) [11], as an architecture description language, to model fault tolerant real-time dynamically reconfigurable systems.

At the design level, as for replication modeling, existing approaches were consisting on manual and explicit redundancy by replicating components, connections and behaviors of AADL components. In the case of a very important number of replicated components or even replicas number, this may cause on one hand the risk of errors and the loss of design time on the other hand. Besides, both the design and the implementation of the software systems are no longer focusing only on functional concerns but also on crosscutting ones.

To solve this problem, we propose in this paper our approach, based on model transformation, to allow us to enrich an AADL model by replication concepts. Our contribution consists on proposing a new technique to design replication using the properties extension provided by AADL in three steps.

The first consists on defining and validating the core model of the system using AADL. This model can be expanded with annexes or properties to have finally the AADL model application.

The second step consists on enriching this model with properties declared within our defined property set, to describe the desired replication mechanism. In fact, we defined `Replication_Properties` as a property set that contains a list of property definitions. These properties describe the adopted replication concepts such as the replication style and the number of replicas. Then, they are applied to AADL specification in order to design the replication of some AADL components at model level.

The final step consists on applying a list of transformation rules to reach an expanded AADL model containing the different replicas. Thus, this approach takes advantage of the extensions provided by the AADL language. It guarantees reducing the complexity of fault tolerant system and gaining the design time based on the separation of concerns and the automatic model transformation.

The three steps will be detailed in the remainder of this paper which is organized as follows: in Sect. 2, we present an overview of the AADL language. Then, we detail the basic concepts of software and hardware FT techniques. We review some related work in Sect. 3. After that, we briefly describe our global approach in Sect. 4 to design both active and passive replication with the property set detailed in Sect. 5. Section 6 describes the established transformation rules. Then, we illustrate the use of our approach by a case study in Sect. 7. Finally, Sect. 8 concludes this paper and gives future work.

2 Background

This section presents an overview about the AADL language. Then, it introduces the basic concepts of software fault tolerance.

2.1 Introduction to the AADL Language

AADL is a standard consisting of both textual and graphical representations with precise execution semantics for embedded software systems. AADL is a typed language providing formal modeling concepts to design the runtime architectures of complex systems and the mapping of software components onto hardware ones through interfaces. This standard defines several categories of components grouped into three subsets:

1. **Software components** including `process`, `thread`, `data`, and `subprogram` components. They may have associated source text specified using property associations. In order to obtain a binary executable image, software source text, coded either in a very-high-level or domain-specific language or in a traditional programming language, can be processed by source text tools.
2. **Hardware or execution platform components** including `device`, `bus`, `memory`, and `processor` components. They represent computing hardware components.
3. **Composite component** including only the `system` component. It represents a composition of software, execution platform, or other system components. System modeling reflects a structure of interacting components organized into a hierarchy.

All hardware, software or composite AADL components of an AADL model correspond to concrete entities that is why AADL is a concrete language. Each of these components can be connected to others through features. These features contain `event` and `data ports`, `subprogram access`, `data access` and `bus access`, among others.

Moreover, AADL can be also used to describe the dynamic behavior of the runtime architecture due to its modes and mode transitions. A mode represents an operational mode state, which manifests itself as an execution platform or an application system configuration. A mode transition consists on changing the system to a different configuration triggered by an event or an event data on ports named in a mode transition or an event raised by the component itself.

Finally, this language can be extended with either properties or annexes. An AADL property provides descriptive information about model entities such as component types, component implementations or subcomponents through a named grouping of property declarations known as *property set*. Thus, AADL offers the possibility to define new properties and property types that can be included in an AADL specification. Therefore, the declaration and use of properties become part of the specification contrary to annexes. Annex libraries enable a designer to extend and customize the AADL core specification with other concepts specified in a language other than AADL. We can for example enrich an AADL specification with the AADL Error Model annex [12] to specify fault tolerance requirements for core components like propagation.

2.2 Software Fault Tolerance

Fault tolerance [1], one of the different means of dependability, is defined as the capability of a system to continue providing offered services even in the presence of errors. Fault tolerance, together with the other means of dependability, address all similar threats that are faults, errors and failures. A **fault** is a physical hardware or software defect causing service degradation. An **error** is an incorrect value causing a system failure. A **failure** manifests a deviation of the system relative to its specification.

The error propagation is a major risk for dependability, in particular fault tolerance. In fact, an error is the manifestation of the fault on the system. When an error is propagated to the service interface and deviates the service from its correct specification, a service failure occurs as a result. The service failure can cause in turn the failure of the whole system. In order to avoid it, software and hardware fault tolerance are accomplished through the following techniques:

1. **Error detection**: consists on detecting error occurrence. It is either concomitant or preemptive.
2. **System recovery**: consists on replacing the erroneous state of the system by another safe state. This technique is based on two mechanisms:

   • *Error handling*, eliminates errors from the state of the system. This is using one of the following techniques or combining some of them in particular situations. The first one, called `Rollback Recovery`, turns up the erroneous state of the system to an earlier saved state. The second, called `Rollforward Recovery` moves the system to a new steady state in order to correct it. The last one is the `compensation` that provides from the beginning enough redundancy of the system to be able to mask erroneous states.

- *Fault handling*, prevents the activation of fault once more. This can be achieved by various ways such as diagnosis, isolation, reconfiguration and reinitialization.

3 Related Work

The literature about fault tolerance techniques used to handle software faults is fairly vast. For example, the authors of [3] gave a survey of software techniques to handle software faults developed in the fault tolerance and the autonomic computing domains. As these techniques are all practically exploiting some form of redundancy, they considered the impact of replication on the software architecture. After that, in order to compare and classify techniques to handle software faults, they proposed a taxonomy based on the nature and use of redundancy in such systems.

Authors in [4] applied redundancy patterns in the architecture design level using Aspect Oriented Paradigm. They focused on the weaving of an original (non-redundant) architecture model with redundancy related design patterns. This approach aims at separating functional and non-functional design. The base model is designed using UML. Then, an aspect model is integrated within the base one using a model weaver. Thus, reusable fault tolerance and redundancy management mechanisms together with their specific analysis sub-models were available in the form of a design pattern library.

Based also on UML designs, authors in [2] propose MARTE-DAM: a profile to support the dependability modeling and quantitative analysis. Unlike several works aiming at extending UML models with dependability annotations, this profile covers different dependability aspects through rich domain models. The defined domain concepts are then mapped to elements of the UML profile. In particular, a redundancy model introduces fault tolerant components which can provide a redundant structure such as variants, deciders (adjudicators), and FT strategies. For performance and dependability analysis and assessment purposes, authors translated the annotated MARTE-DAM into Deterministic and Stochastic Petri Nets (DSPN) models. Even if authors have been proposed the model refinement and dependability assessment, they did not support code generation for MARTE-DAM.

In [9], authors proposed an approach to model and to formally verify replication patterns in the AADL language and then analyze potentially unintended behaviors. This approach is based on designing two AADL models. The first defines the intended behavior in synchronous call sequences and the second describes the replication architecture. A primary-backup replication approach is designed using AADL based on modes and mode transitions. Authors propose two replicas: one primary and one backup. The transitions between primary and backup modes is triggered with an event port. The latter is then connected to a transition controller unit that represents either the human operator or the failure detection module.

In [8], authors gave an example of a primary-backup replication strategy designed with AADL and its behavioral Annex (AADL-BA). They modeled the core system using AADL components and their connections through features. Threads in this case are synchronized using dispatched events. Then, based on AADL-BA, they modeled the automaton showing different states where the application can be blocked to describe the executed call sequences of different threads. They proved also that AADL-BA provides an interesting additional strategy to define critical regions. This work difficulties reside in the design of complex synchronization mechanisms commonly used in distributed system design such as mutual exclusion. Besides, designers who applied this approach must specify both their core system and the replication pattern manually. There are no automatic tasks to help them generating consistent model.

The main difference between the reviewed approaches above and our own is the focus on:

(i) the two replication styles, active and passive one, using the same concepts based on extending the core model with AADL property set

(ii) the automatic code generation of the enriched AADL model following a set of model transformation rules. That is they proposed manual extension of their model integrating active or primary-backup replication style but not both. Also they do not take into consideration the automatic code generation of replication mechanisms.

(iii) the separation of functional and non-functional design as the fault tolerance requirements are specified separately by property set.

4 Proposed Approach

As we have already mentioned, we choose AADL [11] to model fault tolerant real-time dynamically reconfigurable systems. This is for all reasons quoted in Sect. 2.1. We proposed in previous work [6], a development process for the design, implementation and code generation of fault tolerant reconfigurable real-time systems. We used AADL to design not only functional concerns but also crosscutting ones using its different annexes. We decided to use the AADL standard annex E: Error Model Annex [12] to integrate fault tolerant requirements since the design level. In fact, this annex lets us design all types of faults, fault behavior, fault propagation, fault detection and also fault recovery mechanisms. Despite its support of several concepts related to fault tolerance and more generally dependability, this annex does not support the design of replication techniques. It consists of manual redundancy of AADL components, connections and also behaviors. However, the more the replicas or the replicated component are, the more complex and error prone the model is. For that, we decided to propose our own approach in order to help the designer modeling the fault tolerant system easier with integrating replication techniques.

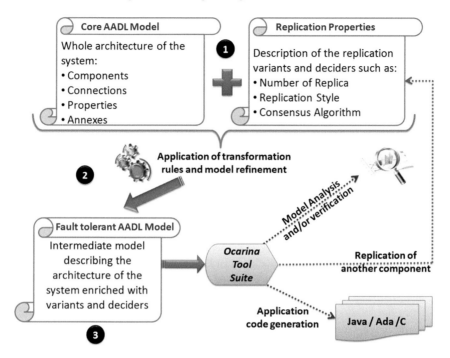

Fig. 1 Replication design process

Our approach takes advantage of the possible extensions of AADL using property sets and annexes. To integrate the model of replication techniques since the design level, we decided to enrich the basic AADL model by a property set that we defined and baptized `Replication_Properties`. It has a set of property declarations and property types describing the adopted replication mechanism as detailed in Sect. 5.

For that, we propose a model driven approach based on AADL model transformation as shown in Fig. 1. Our approach consists on three steps.

We start from a basic (non-redundant) AADL specification model describing an embedded real-time system. We offer the possibility to the designer to enhance his model by predefined properties and annexes. For example, the designer extends his model by clauses from the Error Model Annex for fault tolerance purposes. After that, the designer enriches his AADL model by properties related to the replication strategies through our property set `Replication_Properties`.

After specifying all properties, we validate them before automatically generating an intermediate AADL model enriched with different replicas and deciders. The application of the replication techniques may violate real-time constraints or the original model properties like scheduling or access rights to shared data. Furthermore, if a replicated component of the model is enriched by properties or annexes, then all replicas should inherit them. For example, replicated components may be enriched with the Error Model Annex to model detected and propagated errors and their behaviors in case of errors. So, if a thread is at the same time a replication object and has a particular behavior in case of detected errors, then all generated threads should have the same behavior if the same error is detected. Thus, the replicas are enriched also with the same annexes as well as the replicated one.

For this, we defined a set of transformation rules to govern the model generation process in order to map between the basic AADL model and its extended version with replication concepts. It concerns the generation of either new components (like variants and deciders) or connections between original and generated ones. It ensures also the treatment of the behaviors of the replicas and the decider. These rules are first established manually and then translated into some algorithms implemented from scratch to ensure model transformation. We implemented these rules as an extension of the Ocarina tool suite [14] for these reasons. Ocarina is used to manipulate AADL models. It ensures their syntactic and semantic analysis. In addition, this tool supports scheduling analysis of AADL model with Cheddar [13]. Besides, it offers formal model generation from AADL models with Petri nets. Finally, it allows to generate the code corresponding to the functional part of an AADL specification into **Ada**, **C**, or RTSJ (**Real Time Specification for Java**).

Once the intermediate model is generated and validated, the Ocarina tool suite allows us to generate functional concerns into Ada as it is well adapted to implement real-time embedded systems. For compliance reasons, to generate crosscutting concerns code, we selected AspectAda [5] language. In fact, this language extends the Ada language by aspect concepts. Its design and implementation were defined by the way that it respects real-time constraints. The separation of concerns guarantees better code quality and modularity. It also allows to simplify the validation of either model or code application without effecting each other. Thereby, we can either ensure model analysis or verification of the generated AADL resulting model or generate its corresponding application code (into Ada, C or RTSJ). We can also re-enrich it again by the `Replication_Properties` applied to another component as depicted in Fig. 1.

5 Description of Our Property Set

In the previous section, we have described our approach based on the extension of an AADL model with the `Replication_Properties` property set to enrich it by replication concepts. It contains several properties describing with details the adopted replication mechanism. In this section, we list and explain properties specifically defined for replication purposes.

5.1 Replication Context Description

To describe the context of replication, we defined a property baptized `Description`. The designer gives details about the purpose of replication, its manner or its requirements.

Since this property provides information about the context of replication without any impact on the replication policies, it will be generated as a comment. It is to contribute to the documentation of the generated intermediate model helping the designer understand it.

5.2 Number of Replicas

Using our approach, a single model can support several replicated components. To set the number of replicas that we desire model, we defined a property named `Replica_Number` applied to a given replicated component. This property is set by the designer and corresponds exactly to the number of replicated components at the generated model level. According to [7], the number of replicas (*variants*) depends not only on the FT strategy (RB,[1] NSCP[2] or NVP[3]) (soft or solid) of faults to be tolerated. Thus, we decided to bound the number of replicas through two parameterized constants, `Min_Nbr_Replica` and `Max_Nbr_Replica`, which define respectively the minimal and the maximal number of replicas in the application. In order to give more flexibility to our approach, both constants can be changed by the designer when using our property set.

5.3 Identifiers of Replicas

In order to identify each of the generated replicas, we defined a property named `Replica_Identifiers` composed of a list of string. Each represents an identifier of a generated replica.

5.4 Replication Style

As said above, we are interested in our context to active and passive replication of both hardware and software components. For that, we have defined a property type called `Replication_Types` to describe the adopted style of replication. Such a

---

[1]Recovery Block.

[2]N Self-Checking Programming.

[3]N-Version Programming.

property specifies the associated replication type to a given replicated component. The replication style is then set by the defined property named `Replica_Type`. The value of this property may be either `ACTIVE` or `PASSIVE`. There is no default value for such a property: if this property is not specified, the replication type is undefined and an error is raised to the designer.

5.5 Consensus Algorithm

The consensus (or agreement) algorithm is required in two cases related to the replication style. The first is to elect from one or more secondary copies a new primary copy in case of failure of the current one, in the context of passive replication. The second is to vote between the different replicas in the context of active replication. In both cases, this algorithm can be described via an AADL subprogram component.

The implementation of an AADL subprogram is supplied by the user either as an external file written with another language like **Ada** or **C**, or as an element of the AADL model itself executed by a thread. In order to refer to elements of the AADL model itself, AADL defines two kinds of property types: the `classifier` and the `reference`. For that, to cover all possible cases, we have defined 3 properties: `Consensus_Algorithm_Source_Text`, `Consensus_Algorithm-_Class` and `Consensus_Algorithm_Ref`. These three properties are applied to different kinds of components. They can be applied to features (ports or data access) in the case of active replication of either software or hardware components. In this case, it has a set of outputs that to make decision about same outputs from several replicated components, we apply the consensus algorithm to these features. In case of passive replication, the property describing consensus algorithm must be applied to the replicated component as it is an election algorithm for the decision about the choice of primary copy.

There are no default values for all these properties: if one of them is not specified for a given replicated component, the replication extension is not performed and an error is raised to the designer.

All of these properties contained in the property set `Replication_Properties` were validated by the Ocarina tool suite. Considering it as an AADL specification, this property set was parsed successfully. It was then validated using a set of AADL model examples. In the following, we present a set of transformation rules illustrated by some examples. Some proofs of validation are given and other less important are omitted due to the lack of space.[4]

---

[4]More details about the `Replication_Properties` property set, the transformation algorithms and the case study are available at http://goo.gl/EEQhLK.

6 Model Transformation

As noted previously, we propose in this work the extension of AADL with replication concepts. In the first step, the designer specifies his architectural model. The conceived model is then enriched by a set of properties that we defined in the previous section. After that, based on defined transformation rules, we generate automatically an intermediate AADL model extended with replicas and decider specification. It consists on a direct-manipulation M2M transformation to integrate replication policies to the basic model in order to save effort and reduce errors. Based on automatic generation, we aim at applying a set of transformation rules to get a new enriched model that has to be itself consistent and coherent. The transformation rules, to map the replication concepts into the enriched AADL model, depend on various constraints described in the following subsections.

6.1 Replicated AADL Component Subset

The replication of software component is quite different from hardware or hybrid components. This difference is due to the containment hierarchy of these components and then the possible connections that can be established, the modes clauses that can be or cannot be declared and finally the conjunction with the decider regarding the AADL hierarchy. In addition, by FT communities (for example in [7]), software and hardware fault tolerance architectures and even implementations are not similarly applicable.

For that, we have studied the possibility of replication for each type of component. We support the replication of `threads` and `processes` as software components and `processors` and `devices` as hardware components. We support also the replication of the component `system`. We do not support the replication of `data` and `subprogram` components as we require in this case to apply diversity concepts and not replication ones [3]. In fact, the **diversity** aims at providing the same service through a distinct model and implementation. Replication of identical subprograms does not guarantee better reliability from the treatment viewpoint. For that, we have applied the property declarations defined in the `Replication_Properties` property set into only a subset of the AADL components that we find necessary and consistent.

6.2 Replication Type

The generation of the AADL model enriched by replicas strongly depends on the type of replication defined by the property `Replication_Properties::Replica_Type`.

The adopted replication policies are not the same in the case of active or passive replication.

- **Active Replication**: The generated model contains `Replica_Number` replicas generated inside the same containment hierarchy of the replicated component. Each of them is then connected directly or remotely to a generated or called decider (voter in this case) depending on the property used to specify the consensus algorithm and the type of the replicated component. This replication type distinguishes between software, hardware and system component.
- **Primary-backup Replication**: Unlike active replication, passive one does not distinguish between the different possible types of replication component object. This type of replication, based on the migration between two or more configurations, imposes the generation of `Replica_Number` identical components supporting the dynamic reconfiguration to obey the adaptation needs. To do this, we decided to use the concepts of modes and mode transitions provided by the AADL standard to describe the dynamic behavior of the runtime architecture. Therefore, it is necessary to establish the suitable reconfiguration constraints, responsible of switching between modes, to guarantee that mode transition always bring us to a new safe mode.

6.3 Features of the Replicated Component

The type of features (`ports`, `data access` or `subprogram access`) affects the assumed replication policies. For each feature of replicated component of type `in out` or `out port` or `data access`, we specify its corresponding voter subprogram. That means that the consensus algorithm property is applied to each feature of the replicated component and not to the component itself in the case of active replication unlike the passive replication.

6.4 Consensus Algorithm

The property describing the consensus algorithm has a significant impact on the generated model. The different properties, described in Sect. 5.5, support several configurations of the decider. So, such a property specifies the way that connects replicas to voter even by remote connection in the case of hardware component. For that, we conducted a depth study aiming at discussing all possible cases of active replication of AADL components to tolerate both software and hardware faults.

The description of the consensus algorithm is set through one of the three properties that we have already defined. This property, applied to the replicated component or to each of its features, is then transformed to an AADL subprogram. To be executed, this subprogram must be called by an existent or a generated thread located

Table 1 Generated voter depending on the replicated component type

Replicated component type		Generated voter
Software	**Thread**	– Thread
		– Located at the same containment hierarchy of the replicated component (process)
		– Calls the voter subprogram
	Process	– Process
		– Located at the same containment hierarchy of the replicated component (system)
		– If not specified, runs on the same processor of the original replica
		– Contains itself one voter thread calling the voter subprogram
Hardware or Hybrid		– Thread
		– The voter must be generated inside a software component to be executed. For that, we have to follow the routing connections starting from each feature of the replicated component until reaching the process to which it is linked

itself in a process component. For that, we consider the following different cases summarized in Table 1.

Currently, we are implementing the different transformation algorithms as an extension of the Ocarina tool suite to support replication mechanisms.

7 Case Study

To validate our approach, we describe in this section an example of a simple AADL specification chosen as a case study. This system, presented in Fig. 2, is composed of an AADL system containing three devices describing different sensors and two processes bound onto the same processor. This system is then extended progressively to illustrate our approach. For example, to apply an active replication of the temperature_sensor component described with a device AADL component, we extended its textual description by the lines described in listing 1. We illustrate in this listing the use of the properties defined and explained in previous sections.

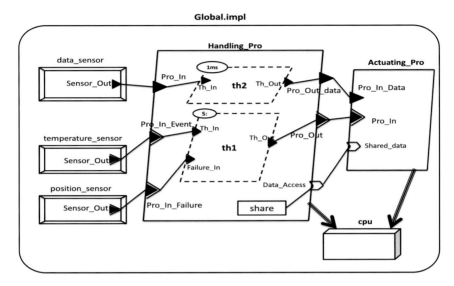

Fig. 2 AADL specification of the basic case study

Listing 1: Replication of the `temperature_Sensor` device component

```
1    system implementation global.impl
2    ...
3    properties
4        Replication_Properties::Description ⇒ "The_temperature_sensor_Replication"
5            applies to temperature_Sensor;
6        Replication_Properties::Replica_Number⇒ 3 applies to temperature_Sensor;
7        Replication_Properties::Replica_Type⇒ ACTIVE applies to temperature_Sensor;
8        Replication_Properties::Replica_Identifiers⇒ ("temp1", "temp2", "temp3")
9            applies to temperature_Sensor;
10       Replication_Properties::Consensus_Algorithm_Source_Text ⇒ "Voting.Do_Vote"
11           applies to temperature_Sensor.sensor_Out;
```

We show in Fig. 3 the generated model after applying the active replication of the temperature sensor described in Fig. 2 and specified by the properties extensions described in listing 1. This involves creating replicas (`temp1`, `temp2` and `temp3`) and voters (thread `Voter`) and establishing the necessary connections between them and original components (`handling_Pro` and `th1`). Generated components inherit automatically the properties applied to the replicated component.

As previously noted, the replication of software component is distinct to the hardware or composite components. For that, we give an other example enriching our case study to illustrate the active replication of a process component. In fact, we extended the generated intermediate model by applying again the replication properties as depicted in listing 2. We show also in this example, the application of the different consensus algorithm into each kind of feature of the replicated process component.

Listing 2: Replication of the `handling_Pro` process component

```
1   system implementation global.impl
2   ...
3   properties
4       Replication_Properties::Description => "Handling_process_replication"
5       applies to handling_Pro;
6       Replication_Properties::Replica_Number => 2 applies to handling_Pro;
7       Replication_Properties::Replica_Type => ACTIVE applies to handling_Pro;
8       Replication_Properties::Replica_Identifiers => ("handling_Pro",
9       "handling_Pro_bis") applies to handling_Pro;
10      Replication_Properties::Consensus_Algorithm_Source_Text => "Voting.Do_Vote"
11      applies to handling_Pro.ProOut_data;
12      Replication_Properties::Consensus_Algorithm_Ref => reference
13      (actuating_Pro.action_Thread.spg_call) applies to handling_Pro.DataAccess;
14      Replication_Properties::Consensus_Algorithm_Class => classifier
15      (Do_treatment) applies to handling_Pro.ProOut;
```

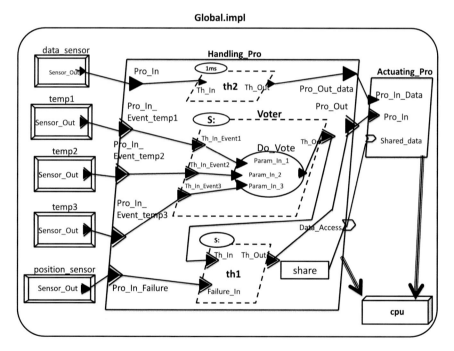

Fig. 3 Active replication of a device component

We do not represent the model resulting from the replication of the process `handling_Pro` due to the lack of space.[5]

The resulting model was validated and instantiated using ocarina. The code application can be after that generated into Ada, C or RTSJ. If desired, other analyses

---

[5]The textual model generated after applying the list of transformation rules of the model shown in Fig. 3 and enriched with properties in listing 2 is available at http://goo.gl/EEQhLK.

can thereby be applied on this resulting model such as schedulability analysis with cheddar or evaluation of the dependability measures with Petri nets.

We deduce from this case study how much the generated model is complicated. Instead of doing it manually, our approach helps the designers to save efforts and reduce the design time and the risk of errors that may appear due to the important number of components and connections. This is by ensuring automatic model transformation using our extension of the Ocarina tool suite.

8 Conclusion and Future Work

We proposed, in this paper, a model level replication approach based on AADL extension to ensure fault tolerance. We took advantage of the extensions possible for AADL language. We defined our property set to support replication concepts since the design level. Then, based on a set of transformation rules, we offer the automatic generation of a new AADL specification enriched with replication concepts. Thus, we help the designer to model his fault tolerant system using AADL easier, at lower costs and more safely.

We have achieved the definition of the transformation rules. We have also integrated the property set `Replication_Properties` into the Ocarina tool support and extended it by the implementation of the transformation rules.

To validate our approach, we tested the model generation for some examples. We gave an example of an AADL specification enriched by active replication of both a device and a process component.

As future work, we aim at accomplishing the implementation of the transformation rules related to the primary-backup replication based on modes and mode transitions. Finally, we aim at formally verifying and proving the correctness of the generated model using a model checker.

References

1. Avizienis, A., Laprie, J.-C., Randell, B., Landwehr, C.: Basic concepts and taxonomy of dependable and secure computing. IEEE Trans. Dependable Secur. Comput. **1**(1), 11–33 (2004)
2. Bernardi, S., Merseguer, J., Petriu, D.: A dependability profile within marte. Softw. Syst. Model. **10**(3), 313–336 (2011)
3. Carzaniga, A., Gorla, A.: and M. Pezz. Handling software faults with redundancy. In: Lemos, R., Fabre, J.-C., Gacek, C., Gadducci, F., Beek, M. (eds.) Architecting Dependable Systems VI. Lecture Notes in Computer Science, vol. 5835, pp. 148–171. Springer, Berlin Heidelberg (2009)
4. Domokos, P., Majzik, I.: Design and analysis of fault tolerant architectures by model weaving. In: International Symposium on High-Assurance Systems Engineering (HASE) (2005)
5. Gabsi, W., Bouaziz, R., Zalila, B.: Towards an aspect oriented language compliant with real time constraints. In: WETICE - AROSA, pp. 68–73. IEEE Computer Society, Hammamet, Tunisia (2013)

6. Gabsi, W., Zalila, B.: Fault tolerance for distributed real time dynamically reconfigurable systems from modeling to implementation. In: WETICE - AROSA, pp. 98–103. IEEE Computer Society, Hammamet, Tunisia (2013)
7. Laprie, J.-C., Béounes, C., Kanoun, K.: Definition and analysis of hardware- and software-fault-tolerant architectures. Computer **23**(7), 39–51 (1990)
8. Lasnier, G., Robert, T., Pautet, L., Kordon, F., Behavioral modular description of fault tolerant distributed systems with aadl behavioral annex. In: NOTERE, pp. 17–24 (2010)
9. Niz, D.D., Feiler, P.H.: Verification of replication architectures in aadl. In: ICECCS, pp. 365–370 (2009)
10. Pinho, L., Vasques, F., Wellings, A.: Replication management in reliable real-time systems. Real-Time Syst. **26**(3), 261–296 (2004)
11. SAE.: Architecture Analysis and Design Language (April 2011)
12. SAE.: Architecture Analysis and Design Language Annex E: Error Model Annex (June 2014)
13. Singhoff, F., Legrand, J., Nana, L., Marcé, L.: Cheddar: a flexible real time scheduling framework. In: International ACM SIGADA Conference, pages 1–8. Atlanta (2004)
14. Vergnaud, T., Zalila, B., Hugues, J.: Ocarina: a Compiler for the AADL. Technical Report, Telecom Paristech - France (2006)

Model Inference of Mobile Applications with Dynamic State Abstraction

Sébastien Salva, Patrice Laurençot and Stassia R. Zafimiharisoa

Abstract We propose an automatic testing method of mobile applications, which also learns formal models expressing navigational paths and application states. We focus on the quality of the models to later perform analysis (verification or test case generation). In this context, our algorithm infers formal and exact models that capture the events applied while testing, the content of the observed screens and the application environment changes. A key feature of the algorithm is that it avoids the state space explosion problem by dynamically constructing state equivalence classes to slice the state space domain of an application in a finite manner and to explore these equivalence classes. We implemented this algorithm on the tool *MCrawlT* that was used for experimentations. The results show that *MCrawlT* achieves significantly better code coverage than several available tools in a given time budget.

Keywords Model inference · Automatic testing · Android applications · State abstraction

1 Introduction

Desktop, Web and more recently mobile applications are becoming increasingly prevalent nowadays and a plethora are now developed for several heterogeneous platforms. All these pieces of software need to be tested to assess the quality of their features in terms of functionalities e.g., conformance, security, performance, etc. Manual testing is the most employed approach for testing them, but manual

S. Salva (✉)
LIMOS CNRS UMR 6158, University of Auvergne, Clermont-Ferrand, France
e-mail: sebastien.salva@udamail.fr

P. Laurençot · S.R. Zafimiharisoa
LIMOS CNRS UMR 6158, Blaise Pascal University, Clermont-Ferrand, France
e-mail: laurencot@isima.fr

S.R. Zafimiharisoa
e-mail: s.zafimiharisoa@openium.fr

© Springer International Publishing Switzerland 2016
R. Lee (ed.), *Software Engineering, Artificial Intelligence, Networking
and Parallel/Distributed Computing 2015*, Studies in Computational Intelligence 612,
DOI 10.1007/978-3-319-23509-7_13

testing is often error-prone and insufficient to achieve high code coverage. These applications share a common feature that can be used for automatic testing: they expose GUIs (Graphical User Interface) for user interaction which can be automatically experimented and explored. Several works already deal with GUI applications testing e.g., desktop applications [9], Web applications [3] or mobile ones [2]. These approaches interact with applications in an attempt to detect bugs and eventually to record models, but all with the same purpose: to obtain good code coverage quickly.

The work, proposed in this paper, falls under this automatic testing category and tackles the testing of mobile applications but also, and above all the learning of models. Our study of model inference techniques has revealed that they often leave aside the notion of correctness of the learned models. This feature is not required for just detecting bug, but is mandatory if models are later used for analysis. Indeed, false models may easily lead to false positives. The quality of the model with regards to its level of abstraction and the amount of information it captures is important as well. Indeed, the more data we collect, the more precise an analysis can be done thereafter. Nevertheless, large amounts of data often lead to large models, up to a state space explosion problem. Based on these observations, we propose an algorithm that aims at learning exact models of mobile applications. We consider the PLTS model (Parameterised Labelled Transition System) to capture the different events made on GUIs. PLTS states also capture all the observed screen contents and notifications about the modifications of the application environment. These notifications signal system events e.g., local database modifications or remote server calls. All this amount of data provide a rich expressiveness that is used while learning the model and that may be later considered for precise model analysis. To avoid a state space explosion, our algorithm dynamically builds state equivalence classes while testing. Each time a new state is discovered, it dynamically re-adjusts the state equivalence relation and classes to limit the state set. These equivalence classes also help recognise similar states that do not require to be explored. Like some available tools [7, 8], our algorithm can also detect application crashes and create test cases for replaying bugs.

We proceed as follows: Sect. 2 briefly presents some related work before introducing an overview of our algorithm that we apply on a straightforward Android application example in Sect. 3. We define the model, the state equivalence relation, and we provide the model inference algorithm in Sect. 4. We give an empirical evaluation on Android applications in Sect. 5 and conclude in Sect. 6.

2 Related Work

Several papers dealing with automatic testing and model generation approaches of black-box systems were issued in the last decade. Due to lack of room, we only present some of them relative to our work. Memon et al. [9] initially presented *GUI Ripper*, a tool for scanning desktop applications. This tool produces event flow graphs and trees showing the GUI execution behaviours. Only the click event can be applied and *GUI Ripper* produces many false event sequences which may need to be weeded out later. Furthermore, the actions provided in the generated models are quite simple (no parameters). Mesbah et al. [10] proposed the tool *Crawljax* specialised in Ajax applications. It produces state machine models to capture the changes of DOM structures of the HTML documents by means of events (click, mouseover,etc.). To avoid the state explosion problem, state abstractions must be given manually to extract a model with a manageable size. Furthermore, the concatenation of identical states proposed in [10] is done in our work by minimisation.

Google's *Monkey* [7] is a random testing tool that is considered as a reference in many papers dealing with Android application automatic testing. However, it cannot simulate complex workloads such as authentication, hence it offers light code coverage in such situations. *Dynodroid* [8] is an extension of Monkey supporting system events. No model is provided. Amalfitano et al. [1] proposed *AndroidRipper*, a crawler for crash testing and for regression test case generation. A simple model, called GUI tree, depicts the observed screens. Then, paths of the tree not terminated by a crash detection, are used to re-generate regression test cases. Yang et al. proposed the tool *Orbit* [12] whose novelty lies in the static analysis of Android application source code to infer the events that can be applied on screens. Then, a classical crawling technique is employed to derive a tree labelled by events. The algorithm implemented in *SwiftHand* [4] is based on the learning algorithm L^* to generate approximate models. The algorithm is composed of a testing engine, which executes applications to check if event sequences meet the model under generation until a counterexample is found. An active learning algorithm repeatedly asks the testing engine observation sequences to infer and eventually regenerate the model w.r.t. all the event and observation sequences.

To prevent from a state space explosion, the approaches [9, 10, 12] require state-abstractions given by users and specified in a high level of abstraction. Choi et al. [4] prefer using the approximate learning algorithm $L*$. These choices are particularly suitable for inferring models for comprehension aid, but these models often are over approximations and given in a high level of abstraction, which may lead to many false positives with test case generation. In this paper, we focus on the inference of exact models. As in [1, 10], we consider the notion of state abstraction that we formally define to limit the state space domain to be explored. But, our algorithm also dynamically re-adjusts state equivalence classes to restrain the exploration and constructs a state abstraction according to the content of the application.

3 Overview

In the following, we present an overview on our model inference algorithm. Before-hand, we give some assumptions on mobile applications considered to design our approach:

Mobile application testing: we consider black-box applications which can be exercised through screens. It is possible to dynamically inspect application states to collect Widget properties. The set of UI events enabled on a screen should be collected as well. If not, Widgets provide enough information (type, etc.) to determine the set of events that may be triggered. Furthermore, any new screen can be observed and inspected (including application crashes). The application environment modifications (databases, network traffic, etc.) can be observed with probes,

Application reset: we assume that mobile applications and their environments (database, remote servers or mocked servers, Operating Systems) can be reset,

Back mechanism availability: several operating systems or applications (Web navigators, etc.) also propose a specialised mechanism, called the *back mechanism* to let users going back to the previous state of an application by undoing its last action. We do not consider that this mechanism is necessarily available and, if available, we assume that it does not always allow to go back to the previous state of an application (wrong implementation, unreachable state, etc.). Most of the other methods assume that the back mechanism always works as expected [1, 8], but this is frequently not the case.

3.1 Terminology

Mobile applications depict screens, which represent application states, the number of states being potentially infinite. Screens are built by application components; here we take back the notation used with Android applications, i.e. *Activities*. The later display screens by instantiating Widgets (buttons, text fields, etc.) which are organised into a tree structure. They also declare the available events that may be triggered by users (click, swipe, etc.). A Widget is characterised by a set of properties (colour, text values, etc.). Hence, one Activity can depict several screens, composed of different Widgets or composed of the same Widgets but having different properties.

Figure 1 depicts the screens of an Android application example used throughout the paper. This application converts colour formats from RGB to HSL (hue-saturation-lightness) and vice-versa by means of two radio buttons $r1$ and $r2$. When the button *Convert* is pressed, the value entered in the blank text field txt is converted and the result appears in the red text field $result$. The chosen colour is also displayed in a colour-box which is depicted at the screen bottom. This application is composed of one Activity which can display an infinite number of screens composed of different text fields values and colour-boxes.

3.2 Algorithm Overview

Figure 2 introduces an overview of our algorithm which is composed of two parts. The algorithm is framed on the task-pool paradigm (Fig. 2a). Tasks are placed into the task-pool, implemented as an ordered list, and each can be executed in parallel. A task $Explore(q, p)$ corresponds to one screen to explore. A screen is transcribed

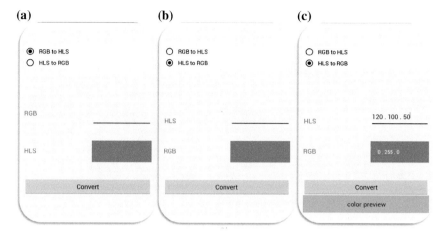

Fig. 1 Colour converter android application

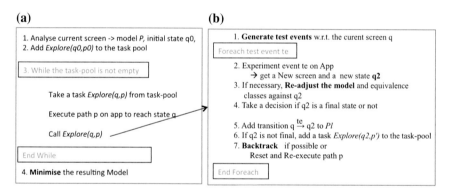

Fig. 2 Overview of the model inference algorithm. **a** Application exploration algorithm. **b** Explore procedure

by the state q gathering all the Widget properties composing the screen and p is a path allowing to reach q from the initial state q_0. When there is no more task to do, the exploration implicitly ends. The resulting model is then minimised to be more readable.

The exploration of one state (Fig. 2b) is done by the Explore procedure. A set of test events (parameter values combined with an event set), which match the current application state, is firstly generated. The current screen is experimented with every test event to produce new screens. However, this step may lead to an infinite set of states to explore. To avoid this well-known issue, the algorithm slices the state space domain into a finite state equivalence class set by means of an equivalence relation (defined in Sect. 4.1). A state which belongs to a previously discovered equivalence class is marked as final otherwise it has to be explored. Intuitively, for every new built state q_2 (step 2), the algorithm eventually readjusts the state abstraction to limit the

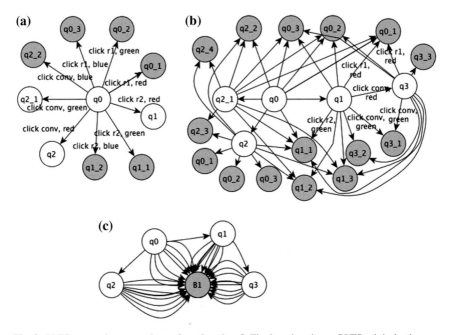

Fig. 3 PLTS generation example. **a** q0 exploration. **b** Final exploration. **c** PLTS minimisation

state set size (step 3). It scans the detected equivalence classes and checks if some of them (three or more in the algorithm) are different only on account of one Widget property. If so, it has detected a Widget property which may lead to the construction of several equivalence classes and states to explore (up to an infinite set of states). Consequently, it readjusts the equivalence relation, classes and the model by masking this Widget property. This means that this property is no more taken into account for the equivalence class computation. Therefore, the new state q_2 belongs automatically to an already discovered equivalence class and so, it will not be explored. No new equivalence class is built either. Then, the algorithm checks if new states have to be explored (step 4). Finally, the algorithm tries to backtrack the application to go back to its previous state by undoing the previous action. If it doesn't work, the application and its environment (OS, databases, etc.) are reset and the previous path p is used to reach the state which is currently under exploration.

Figure 3 illustrates with simplified graphs (no PLTSs) how the algorithm works on the example of Fig. 1. For simplicity, only three values are considered for testing: colour red (rgb=255,0,0 or hsv=0,100,50), green and blue. We also assume that these values are always used for testing in the same order. The equivalence relation is: *two states are equivalent if they have the same Widget properties, except those related to text field values*. These last properties are usually not considered for conceiving state abstractions since these often lead to a large potentially infinite set of states. Furthermore, if a Widget property takes more than two values in the different equivalence classes then the relation has to be re-adjusted.

1. Initially, we have a state $q0$ which corresponds to the beginning of the application (Fig. 1a) and the corresponding equivalence class $[q0]$. A list of test events is generated from $q0$: the events click on the radio-buttons $r1$, $r2$ or on the button *conv* are combined with the colour values that are injected into the blank text field *txt*. The application is firstly experimented with the click on $r1$ and with the red colour value. This produces a new screen and a state $q0\_1$ which belongs to the same equivalence class $[q0]$ because only the text field *txt* is modified. This new state is marked as final (in grey in Fig. 3) and is not explored. The application is backtracked to return to $q0$. With the other colours, we also reach final states $q0\_2$ and $q0\_3$ which are marked as final because they belong to the same equivalence class $[q0]$.

   Then, the radio-button $r2$ is clicked, with the red colour. We obtain a new state $q1$ (Fig. 1b) and a new equivalence class $[q1]$ since $r2$ is now enabled. Therefore, we get a new task $Explore(q1, p')$. Once more, the application is backtracked. When using the other colour values, we obtain the states $q1\_1$ and $q1\_2$ that are marked as final since these belong to $[q1]$. No task is created.

   When the *conv* button is clicked, a value appears in the text field *result* and a colour is depicted in the colour-box (Fig. 1c). We obtain a state $q2$, which has to be explored, and a new equivalence class $[q2]$. Next, *conv* is clicked with the green colour. The state $q2\_1$ is built with a new equivalence class $[q2\_1]$. This state is not marked as final since the colour-box displays a new colour. This process should continue for every colour and in particular with the blue one, which produces the state $q2\_2$. A state space explosion may happens here. But the algorithm detects that three equivalence classes are different only on account of the same property *colourbox.colour*. The algorithm readjusts the equivalence relation to limit the state set size. Intuitively, the equivalence relation becomes *two states are equivalent if they have the same Widget properties, except those related to text field values and to the colourbox.colour property*. Then, it updates states and equivalence classes to match this new relation. As a consequence, $[q2]$, $[q2\_1]$ and $q2\_2$ are now merged into $[q2]$. The new state $q2\_2$ now belongs to an existing equivalence class and is hence marked as final. The first task $Explore(q0, p)$ is finished and we obtain the graph depicted in Fig. 3a,

2. We assume that the task $Explore(q1, q0 \xrightarrow{click\ r2, txt=red} q1)$ is picked out to explore $q1$. A list of test event, which is the same as previously is constructed. From the state $q1$, when the button *conv* is clicked with the red colour value, a new state $q3$ is added because the colour box appears. When *conv* is clicked with other colour values and events we only obtain final states, since they belong to previously discovered equivalence classes,

3. the same reasoning is followed on states $q2$, $q2\_1$ and $q3$, but only final states are added (no task). We obtain the PLTS of Fig. 3b,

4. the task-pool is empty. The PLTS is finally minimised [6]. Here, the final states are merged to one unique state as illustrated in Fig. 3c.

In this short example, we have shown that our algorithm avoids the state explosion problem and ends once at least one state of all the detected equivalence classes is explored. In the following, we describe formally the model, the equivalence relation, and the algorithm.

4 Model Inference Algorithm

4.1 Mobile Application Modelling with PLTS

We use PLTSs as models for representing mobile applications. A PLTS is a kind of state machine extended with variables and guards on transitions. Beforehand, we assume that there exist a domain of values denoted D and a variable set X taking values in D. The assignment of variables in $Y \subseteq X$ to elements of D is denoted with a mapping $\alpha : Y \rightarrow D$. We denote D_Y the assignment set over Y.

Definition 1 (*PLTS*) A PLTS (Parameterised Labelled Transition System) is a tuple $< V, I, Q, q0, \Sigma, \rightarrow >$ where:

- $V \subseteq X$ is the finite set of variables, $I \subseteq X$ is the finite set of parameters used with actions,
- Q is the finite set of states, such that a state $q \in Q$ is an assignment over D_V, q_0 is the initial state composed of the initial condition D_{V0},
- Σ is the finite set of valued actions $a(\alpha)$ with $\alpha \subseteq D_I$,
- $\rightarrow \subseteq Q \times \Sigma \times Q$ is the transition relation. A transition $(q, a(\alpha), q')$ is also denoted $q \xrightarrow{a(\alpha)} q'$.

The behaviour of a PLTS P is characterised by its sequences of valued actions starting from its initial state $q0$. These sequences are also called the traces of P:

Definition 2 (*PLTS Traces*) Let $P =< V, I, Q, q_0, \Sigma, \rightarrow >$ be a PLTS. $Traces(P)$ $= Traces(q_0) = \{a_1(\alpha_1)...a_n(\alpha_n) \mid \exists q_1, ...q_n, q0 \xrightarrow{a_1(\alpha_1)} q_1...q_{n-1} \xrightarrow{a_n(\alpha_n)} q_n \in (\rightarrow)^*\}$.

We model mobile application behaviours with PLTSs by encoding (UI) events with actions. We also store the properties collected from screens (Widget properties) and notifications about the application environment changes in states with variable assignments:

UI Events Representation

We interact with mobile applications by means of events, e.g., a click on a button, and by entering values into editable Widgets. We capture such events with PLTS actions of the form $event(\alpha)$ with $\alpha = \{widget := w, w_1 := val_1, ..., w_n := val_n\}$

an assignment over D_I; the parameter *widget* denotes the Widget name on which is applied the event and the remaining variables are assignments of Widget properties. We denote the triggering of the back mechanism with the action $back(\alpha)$ with α an empty assignment.

Application State Representation

We specialise PLTS states to store the content of screens (Widget properties) in such a way as to facilitate the construction of equivalence classes of states. We split the set of Widget properties into two categories: we gathers in the set W the Widget properties that indicate a strong application behaviour modification and that take only few values e.g., Widget visibility, size, etc. The others that usually take a lot of different values such as the properties about text field values, are placed into W^c. This separation affects the state representation: we denote wp the assignment composed of properties in W, while the assignment wo is composed of the other Widget properties. A PLTS state q is then a specific assignment of the form $act \cup wp \cup wo \cup env \cup end$ where:

- act is an assignment returning an Activity name,
- (wp, wo) are two Widget property assignments. The union of wp and wo gives all the property values of an application screen displayed by act.
- env is a boolean assignment indicating whether the application environment has been modified,
- end is a boolean assignment marking a state as final or not.

For readability, a state $q = act \cup wp \cup wo \cup env \cup end$ is denoted (act, wp, wo, env, end). This state structure eases the definition of the state equivalence relation given below:

Definition 3 (*State equivalence relation*) Let $P =< V, I, Q, q0, \Sigma, \rightarrow>$ be a PLTS and for $i = (1, 2)$ let $q_i = (act_i, wp_i, wo_i, env_i, end_i)$, be two states in Q. We say that q_1 is equivalent to q_2, denoted $q_1 \sim q_2$ iff $act_1 = act_2, wp_1 = wp_2$ and $env_1 = env_2$. $[q]$ denotes the equivalence class of equivalent states of q. $Q/\sim$ stands for the set of all equivalence classes in Q.

This definition gives a very adaptable state equivalence relation whose meaning can be modified by altering the assignments wp. If we take back our example, one can consider two states $q1 \nsim q2$ which are different only because they do not include the same assignments of the Widget property *colourbox.colour* ($act_1 = act_2$, $env_1 = env_2$, but $wp_1 \neq wp_2$). We have two equivalence classes $[q_1], [q_2]$. The equivalence relation is adaptable in the sense that wp_i can be changed as follows: if we consider that *colourbox.colour* takes too much values and implies to much different equivalence classes, *colourbox.colour* can be shifted from wp_i to wo_i ($i = 1, 2$)

in states. We obtain $wp_1 = wp_2$, $q_1 \sim q_2$ and only one equivalence class $[q_1]$. Intuitively, our algorithm uses this adjustment process to dynamically reduce the equivalence class domain and the state exploration according to the screen content.

4.2 Model Inference Algorithm

In this section, we describe more precisely the Explore procedure (second part of the overview in Fig. 2) whose pseudo code is given in Algorithm 1. Due to lack of room, the task-pool management algorithm can be found in [11].

As stated above, this procedure aims at visiting one state to augment the PLTS under construction, denoted P, with new transitions and states and to eventually produce new tasks $Explore(q, p)$ added to the task-pool. Its steps are explained below:

- Test data generation and execution (lines 4–11): the current screen is analysed to generate a set of events expressing how to complete Widgets with values and to trigger an event. In short, our algorithm generates a set of events of the form $\{event(\alpha) \mid event$ is an event, α is an assignment$\}$. It starts collecting the events that may be applied on the different Widgets of the current screen. Then, it constructs assignments of the form $w_1 = v1 \wedge \dots \wedge w_n = vn$, with $(w_1, \dots, w_n)$ the list of editable Widget properties found on the screen and $(v_1, \dots, v_n)$, a list of test values. Instead of only using random values, we propose to use several data sets: a set $User$ gathering values manually chosen such as logins and passwords, a set RV composed of values well known for detecting bugs e.g., String values like "&", "", or null, and of random values. A last set, denoted $Fakedata$, is composed of fake user identities. Furthermore, we adopted a Pairwise technique [5] to derive a set of assignment tuples over these data sets. Assuming that errors can be revealed by modifying pairs of variables, this technique strongly reduces the coverage of variable domains by constructing discrete combinations for pair of parameters only. Then, each $event(\alpha)$ is applied on the current screen to produce new ones (application crash included). Each screen is analysed to retrieve Widget properties and the activity which produces this screen. Probes are requested to detect if the application environment were modified. These data are formalised by the state q_2,
- Model readjustment: the Explore procedure now checks whether the re-adjustment of P and of the state equivalence classes is required (lines 9–12). We denote $C^{Wprop}(Q/\sim)$ the number of assignments of the same Widget property $Wprop$ found in the set of equivalence classes $Q/\sim$. $C^{Wprop}(Q/\sim m) = card(\{\alpha = (Wprop := val) \mid [q] \in Q/\sim, q = (act, wp, wo, env, end), \alpha \in wp\})$. For each assignment $\alpha = (Wprop := val)$ in wp_2, we check how much values the Widget property $Wprop$ takes in the equivalence classes: if $Wprop$ takes more than 2 values in $Q/\sim$ (if $card(C^{Wprop})(Q/\sim) > 2$), then we re-adjust the state representation. In every state $q = (act, wp, wo, env, end)$ of $Q \cup \{q_2\}$, the

assignments of the form $(Wprop := val)$ are shifted from wp to wo (procedure Readjust in Algorithm 1 line 11). The equivalence classes are also transformed in accordance (procedure Readjust line 12),

- PLTS completion: a new transition $q \xrightarrow{event(\alpha)} q_2$ is added to the PLTS P (lines 13-20). q_2 is marked as final if q_2 belongs to an existing equivalence class. Otherwise (line 17), q_2 has the assignment $(end := false)$ and a new task $Explore(q_2, p')$ is added to the task pool. Since the algorithm is highly parallelisable, we use critical sections to modify the PLTS P (which is shared among threads),
- Application backtracking: to apply the next event, the Explore procedure calls the Backtrack one (line 21) to reach the previous screen and state q. Its algorithm is given in Algorithm 2. Here the notion of application environment really makes a difference to achieve an exact model: if the current state q_2 has an assignment $(env := false)$, its reflects the fact that the application environment has not be modified, therefore the Backtrack procedure calls the back mechanism to undo the most recent action (if available). We observe a new screen and check whether it is equivalent to the previous screen stored in q (we compare their Widget properties). Otherwise, the application and its environment are reset and we re-execute the path p to reach the state q (Algorithm 2, line 7) (here, we assume that the application is deterministic though).

Algorithm 1: Explore Procedure

---

1 Procedure $Explore(q, p)$;
2 $Events = GenEvents$, analyse the current screen to generate the set of events $\{event(\alpha) \mid event$ is an event, α is an assignment$\}$;
3 **foreach** $event(\alpha) \in Events$ **do**
4   Experiment $event(\alpha)$ on $App \rightarrow$ new screen $Inew$;
5   Analyse $Inew \rightarrow$ assignments act_2, wp_2, wo_2;
6   Analyse the application environment $\rightarrow env_2$;
7     $q_2 = (act_2, wp_2, wo_2, env_2, end := null)$;
8   **foreach** $\alpha = \{Wprop := val\} \in wp_2$ **do**
9   **if** $card(C^{Wprop}(Q/\sim) \cup \{\alpha\}) > 2$ **then**
10             $Readjust(Q \cup \{q_2\}, Wprop)$;
11             $Readjust(Q/\sim, Wprop)$;
12   **if** $Inew$ reflects a crash or there exists $[q'] \in Q/\sim$ such that $q_2 \in [q']$ **then**
13         $\{$Add a transition $q \xrightarrow{event(\alpha)} q_2 = (act_2, wp_2, wo_2, env_2, end := true)$ to $\rightarrow_P$;
14         $\}$ (in critical section)
15   **else**
16         $\{$Add a transition $t = q \xrightarrow{event(\alpha)} q_2 = (act_2, wp_2, wo_2, env_2, end := false)$ to $\rightarrow_P$;
17         $Q/\sim = Q/\sim \cup \{[q_2]\}$;
18   Add the task $(Explore(q_2, p.t))$ to the task-pool;
19         $\}$ (in critical section)
20     $Backtrack(q_2, q, p)$;

4.3 PLTS Minimisation

Our algorithm performs a minimisation on the first generated PLTS to achieve a more readable model. We have chosen a bisimulation minimisation technique since this one still preserves the functional behaviours represented in the original model while reducing the state space domain. A detailed algorithm can be found in [6]. In short, this algorithm constructs sets (blocks) of states that are bisimilar equivalent (any action from one of them can be matched by the same action from the other and the arrival states are again bisimilar). Figure 3c depicts the (simplified) minimised PLTS of the application example. Here, final states are aggregated into one block of states.

Algorithm 2: Backtrack procedure Algorithm

---

1 Procedure
  $Backtrack(q_2 = (act2, wp2, wo2, env2, end2), q = (act, wp, wo, env, end), p)$;

2 **if** $env2 = (env := false)$ *and the back mechanism is available* **then**
3 | Call the back mechanism → screen $INew$;
4 | Analyse $Inew$ → assignments rc', wp', wo';
5 | Analyse the application environment → env';
6 | **if** $act \neq act'$ *or* $wp \neq wp'$ *or* $wo \neq wo'$ *or* $env \neq env'$ **then**
7 | Reset and Execute App by covering the actions of p;
8 | **else**
9 | Add a transition $t = q_2 \xrightarrow{back(\alpha)} q$ to $\rightarrow_{Tree}$;

10 **else**
11 | Reset and Execute App by covering the actions of p;

4.4 Algorithm Correctness, Complexity and Termination

We express the correctness of our model inference method in term of trace equivalence between the inferred PLTS and the traces of the application under test:

Proposition 1 *Let P be a PLTS constructed with our model inference algorithm from a deterministic mobile application App. We have $Traces(P) \subseteq Traces(App)$.*

The proof is given in [11]. Intuitively, our algorithm constructs a PLTS P with these steps:

1. *Generation of PLTS:* from a given state q, every new event applied on the application is modelled with a unique transition whose arrival state q_2 is new or final. We do not merge states and hence we construct a PLTS P,

2. *Correct use of the back mechanism:* we call this mechanism with care: it is called only if the environment of the application (databases, remote servers, etc.) were not modified with the execution of the last action. Indeed, if we apply the back mechanism even so, we necessarily reach a new state since the application environment is modified. Secondly, we check if the state of the application obtained after the call of the back mechanism is really the previous state of the application. If one of these conditions is not met, we reset the application and its environment and we re-execute the path p to reach the state q,
3. *Minimisation with trace equivalence:* we apply a bisimulation minimisation technique to produce a PTLS MP from P such that the two PLTS are bisimilar and consequently trace equivalent as well.

Complexity and termination of the Algorithm: our algorithm builds at most $2*2^n$ equivalence classes, with n the number of Widget properties in W. In short, we can have two different ($env := true$, $env := false$) and m^n different assignments over W if m is the maximum number of values that any Widget property can take. Nonetheless, when a property of W takes more than two values, our algorithm shifts it from the assignment wp to wo in states. Furthermore, since we explore one state per equivalence class, the algorithm ends and we have $2*2^n$ equivalence classes and not final states. We also have at most nm transitions (Pairwise testing [5]) for each. If N and M stand for the number of not final states and transitions, the whole algorithm has a complexity proportional to $\mathcal{O}(M + N + MN + Mlog(N))$. Indeed, the Explore procedure covers every transition twice (one time to execute the event and one time to go back to the previous state) and every not final state is processed once. But, sometimes the back mechanism is not available. In this situation, the application is reset to go back to a state q by executing the events of a path p at worst composed of M transitions. In the worst case, this step is done for every state with a complexity proportional to NM. Furthermore, the minimisation procedure has a complexity proportional to $\mathcal{O}(Mlog(N))$ [6].

5 Empirical Evaluation

We present here some experimentations on Android applications to answer on the following questions: does the algorithm offer good code coverage in a reasonable time delay? How are the models in terms of size and quality for analysis?

We have implemented our algorithm in a tool called *MCrawlT* (Mobile Crawler Tool [1]). It takes packaged applications or source projects and user data e.g., logins and passwords required for the application execution. *MCrawlT* is based on the testing framework *Robotium* [2] which retrieves the Widget properties of a screen and simulates events.

---

[1] available here https://github.com/statops/mcrawlert.git.
[2] https://code.google.com/p/robotium/.

To avoid any bias, we compare the effectiveness of *MCrawlT* with the following available tools, *Monkey* [7] and *Dynodroid* [8], on applications taken as reference in the papers [4, 8, 9, 12] and whose source code is available (30 applications). The results of some other tools *Orbit* [12], *Guitar* [9] and *Swifthand* [4] are taken from the papers. It is important to note that *Monkey* is taken as a reference in most of the papers dealing with Android testing. Thereby, our results can be compared with other studies related to Android testing.

Code coverage: Table 1 reports the percentages of code coverage obtained with the different tools on 30 applications with a time budget of three hours. If we do a side by side comparison of *MCrawlT* with the other tools, we observe that *Monkey* provides better code coverage for 8 applications, *SwiftHand* for 2 and *Dynodroid* for 5. In comparison to all the tools together, *MCrawlT* provides better code coverage for 20 applications, the coverage difference being higher than 5 % with 13 applications. These results show that *MCrawlT* gives better code coverage than the other tools and even offers good results against all the tools together on half the applications with comparable execution times. Table 1 also reveals that the obtained code coverage percentage is between 25 and 96 %. We manually analysed the 8 applications which yield the less good results with *MCrawlT* to identify the underlying causes behind low coverage. This can be explained at least by these ways:

- Specific functionalities and unreachable code: several applications are incompletely covered either on account of unused code parts (libraries, packages, etc.) that are not called, or on account of functionalities difficult to start automatically,
- Unsupported events: several applications e.g., Nectdroid, Multism, Acal or Alogcat chosen for experimentation with *Dynodroid* take UI events as inputs but also system events such as *Android broadcast messages*. Our tool does not support these events yet. Moreover, *MCrawlT* only supports the event list also supported by the testing tool *Robotium* (viz. click and scroll). The long click is thus not supported but is used in some applications (Mininote and Contactmanager). In contrast, *Orbit* supports this event and therefore offers a better code coverage with the application Contactmanager.

Quality and size of the models: Table 2 finally shows the number of states obtained with *MCrawlT*, *Orbit* [12] and *SwitHand* [4] since they produce models as well. Before minimisation, our tool generates larger and tacitly less comprehensive models than those obtained with *Orbit*. In term of quality of the learned models, we do not produce extrapolated models and we believe that those generated by *MCrawlT* offer more testing capabilities. Indeed, these models include states which store all the observed Widget properties (colours, texts, etc.) and notifications about the application environment changes. We have precisely chosen this feature to later perform test case generation. For instance, with this amount of information, we can construct test cases to apply events and to check the content of the resulting screen but also if remote servers are called, etc. Both *Orbit* and *SwiftHand* only store UI events. After minimisation, we obtain more compact and readable models whose sizes are comparable to the sizes of the models obtained with *Orbit*. This tends to show that our approach of producing larger but more detailed models that are after minimised,

Table 1 Code coverage (in %)

Applications	Monkey	Orbit	Guitar	MCrawlT	SwifHand	Dynodroid
NotePad	60	82		88		crash
Tippy_TipperV1	41	78		79		48
ToDoManager	71	75	71	81		34
OpenManager	29	63		65		crash
HelloAUT	71	86	51	96		76
TomDroid	46	70		76		42
ContactManager	53	91	71	68		28
Aardict	52	65		67		51
Musicnote	69			81	72.2	47
Explorer	58			74	74	crash
Myexpense	25			61	41.8	40
Anynemo	61			54	52.9	crash
Whohas	58			95	59.3	65
Mininote	42			26	34	39
Weight	51			34	62	56
Tippy_TipperV2	49			74	68	12
Sanity	8			26	19.6	1
Nectdroid	70.7			54		68.6
Alogcat	66.6			66		67.2
ACal	14			46		23
Anycut	67			71		69.7
Mirrored	63			76		60
Jamendo	64			46		3.9
Netcounter	47			56		70
Multisms	65			73		77
Alarm	77			72		55
Bomber	79			75		70
Adsdroid	72			83		80
Aagtl	18			25		17
PasswordFor Android	58			61		58

only offer advantages for model inference. In addition, *MCrawlT* constructs storyboards from these minimized models by replacing states with screen-shots of the application.

All these experimental results on real applications tend to show that our tool is effective and can be used in practice since it produces equivalent or higher code coverages than the other tools.

Table 2 Inferred model size

Applications	#PLTS states (MCrawlT)	#states after minimisation (MCrawlT)	#states (Orbit)	#states (SwiftHand)
NotePad	13	8	7	
Tippy_TipperV1	37	18	9	
ToDoManager	6	2		
OpenManager	31	12	20	
HelloAUT	8	5	8	
TomDroid	12	6	9	
ContactManager	5	4	5	
Sanity	31	24		78
Musicnote	41	23		46
Explorer	96	74		195
Myexpense	52	37		149
Anynemo	139	106		169
Whohas	36	11		97
Mininote	45	19		169
Tippy_TipperV2	54	26		71
Weight	69	23		109

6 Conclusion

In this paper, we present an algorithm, which infers PLTS models from mobile applications. It constructs PLTSs that capture events and all the Widget properties extracted from the observed screens. Despite the huge amount of collected data, we avoid the state space explosion problem by using an equivalence relation and classes that are dynamically re-adjusted all along the algorithm execution with regards to the screen content. Our experimental results show that our algorithm offers good code coverage quickly and can be used in practice. Furthermore, the generated models can be reused for precise model analysis. An immediate line of future work would be to apply this kind of algorithm for security breach detection. The exploration could be specialised to target some specific application parts (login step, etc.). Then, test cases could be automatically generated from test patterns to further explore specific states with the purpose of improving detection.

References

1. Amalfitano, D., Fasolino, A., Tramontana, P.: A gui crawling-based technique for android mobile application testing. In: Software Testing, Verification and Validation Workshops (ICSTW), 2011 IEEE Fourth International Conference on, pp. 252–261 (2011). doi:10.1109/ICSTW.2011.77
2. Anand, S., Naik, M., Harrold, M.J., Yang, H.: Automated concolic testing of smartphone apps. In: Proceedings of the ACM SIGSOFT 20th International Symposium on the Foundations of Software Engineering, FSE '12, pp. 1–11. ACM, New York, NY, USA (2012). doi:10.1145/2393596.2393666
3. Artzi, S., Kiezun, A., Dolby, J., Tip, F., Dig, D., Paradkar, A., Ernst, M.: Finding bugs in web applications using dynamic test generation and explicit-state model checking. IEEE Trans. Softw. Eng. **36**(4), 474–494 (2010). doi:10.1109/TSE.2010.31
4. Choi, W., Necula, G., Sen, K.: Guided gui testing of android apps with minimal restart and approximate learning. SIGPLAN Not. 48(10), 623–640 (2013). doi:10.1145/2544173.2509552, http://doi.acm.org/10.1145/2544173.2509552
5. Cohen, M.B., Gibbons, P.B., Mugridge, W.B., Colbourn, C.J.: Constructing test suites for inter-action testing. In: Proceeding of the 25th International Conference on Software Engineering, pp. 38–48 (2003)
6. Fernandez, J.C.: An implementation of an efficient algorithm for bisimulation equivalence. Sci. Comput. Programm. **13**, 13–219 (1989)
7. Google: Ui/application exerciser Monkey. http://developer.android.com/tools/help/monkey.html, Last Accessed Jan 2015
8. Machiry, A., Tahiliani, R., Naik, M.: Dynodroid: An input generation system for android apps. In: Proceedings of the 2013 9th Joint Meeting on Foundations of Software Engineering, ESEC/FSE 2013, pp. 224–234. ACM, New York, NY, USA (2013). doi:10.1145/2491411.2491450
9. Memon, A., Banerjee, I., Nagarajan, A.: Gui ripping: Reverse engineering of graphical user interfaces for testing. In: Proceedings of the 10th Working Conference on Reverse Engineering, WCRE '03, pp. 260–269. IEEE Computer Society, Washington, DC, USA (2003)
10. Mesbah, A., van Deursen, A., Lenselink, S.: Crawling Ajax-based web applications through dynamic analysis of user interface state changes. ACM Trans. Web (TWEB) **6**(1), 1–30 (2012)
11. Salva, S., Laurençot, P., Zafimiharisoa, S.R.: Model inference of mobile applications with dynamic state abstraction. Techinal Report, LIMOS (2015). http://sebastien.salva.free.fr/RR-15-01.pdf
12. Yang, W., Prasad, M.R., Xie, T.: A grey-box approach for automated gui-model generation of mobile applications. In: Proceedings of the 16th international conference on Fundamental Approaches to Software Engineering, FASE'13, pp. 250–265. Springer, Berlin (2013). doi:10.1007/978-3-642-37057-1_19

Automatic Generation of S-LAM Descriptions from UML/MARTE for the DSE of Massively Parallel Embedded Systems

Manel Ammar, Mouna Baklouti, Maxime Pelcat,
Karol Desnos and Mohamed Abid

Abstract Massively Parallel Multi-Processors System-on-Chip (MP2SoC) architectures require efficient programming models and tools to deal with the massive
parallelism present within the architecture. In this paper, we propose a tool which
automates the generation of the System-Level Architecture Model (S-LAM) from a
Unified Modeling Language-based (UML) model annotated with the Modeling and
Analysis of Real-Time and Embedded Systems (MARTE) profile. The S-LAM-based
description of the MP2SoC architecture is conformed to the IP-XACT standard. The
integration of our generator within a co-design framework provides the specification
of the whole MP2SoC system using UML and MARTE. Then, gradual refinements
allow the execution of a rapid prototyping process.

1 Introduction

Recent trends in High-Performance Computing (HPC) architectures show that, due to
the end of processor frequency scaling, performance increases are mostly gained by
employing more processor cores [1]. This trend draws attention to the effectiveness
of Massively Parallel Multi-Processors System-on-Chip (MP2SoC) architectures in
the HPC domain. Designers of high performance MP2SoC are facing many critical
design challenges including:

M. Ammar (✉) · M. Baklouti · M. Abid
CES Laboratory, National Engineering School of Sfax, Sfax, Tunisia
e-mail: manel.ammar@ceslab.org

M. Pelcat · K. Desnos
IETR, INSA Rennes, CNRS UMR 6164, UEB, Rennes, France
e-mail: mpelcat@insa-rennes.fr

© Springer International Publishing Switzerland 2016   195
R. Lee (ed.), *Software Engineering, Artificial Intelligence, Networking*
and Parallel/Distributed Computing 2015, Studies in Computational Intelligence 612,
DOI 10.1007/978-3-319-23509-7_14

1.1 Raising the Level of Abstraction of the Specification

The raising complexity of embedded systems creates a need for intensive specification task. In the history of design flows, changes in design productivity were
always related to raising the level of abstraction in design entry. In the 1970s, the
highest level of abstraction was a transistor schematic. 10 years later, design entry
had moved up from transistors to gates. Then, with the appearance of Hardware
Description Languages (HDL) other levels of abstraction were proposed including
the Register-Transfer Level (RTL) and the behavioral level. In the beginning of the
2000s, and with the emergence of new languages (mainly SystemC) for the description of systems, a higher level of abstraction was created named the system-level.
Current research targeting the Model Driven Engineering (MDE) methodology [2]
shows the effectiveness of this methodology in the domain of System-on-Chip (SoC)
design. Describing complex systems using models, which is the primary issue of
MDE, leads to the creation of a higher level-of-abstraction: the model level. This
level is mainly based on the Unified Modeling Language (UML) [3] and a domain-
specific profile dealing with a specific type of systems: embedded systems.

1.2 Reusing IP Blocks

Historically, design reuse has proven its utility in the SoC design field as system
complexity continuously increases [4]. However, there is one important challenge
in adopting this methodology: the lack of formal characterization of platforms. As
a result, platforms should be formally defined in terms of semantics to facilitate
verification, automatic design, reuse and interoperability between Electronic Design
Automation (EDA) tools. IP-XACT [5] was created to face this challenge. It describes
electronic components and their designs in an Extensible Markup Language (XML)
format that facilitates exchanging IPs between different EDA tools for complex SoC
design. IP-XACT was standardized by the SPIRIT Consortium.

1.3 Building Well Structured Methodologies

Methods and tools used in the specification and design space exploration of HPC
architectures aim at managing the increasing complexity of hardware architectures
specification task while promoting IP reuse through the IP-XACT standard. Current
hardware specification efforts within the MDE community can be summarized in
two key points:

- Modeling IP-XACT designs in UML and annotating models with IP-XACT specific stereotypes
- Applying UML as high-level specification methodology and link it with IP-XACT in a lower-level of abstraction using MDE transformation rules

The work presented in this paper is an effort towards the second key point. Actually, we propose a new approach that takes advantage from UML as high-level modeling language combined with the Modeling and Analysis of Real-Time and Embedded Systems (MARTE) profile [6] and introduces another level that facilitates IP integration, architecture generation and system analysis. This level is based on the System-Level Architecture Model (S-LAM) [7] which conforms to the IP-XACT standard. S-LAM proposes a simple description of MP2SoC architectures at system-level while reducing the architecture simulation complexity. This paper presents the MARTE to S-LAM generator, able to generate from a UML/MARTE description of the MP2SoC architecture, the corresponding S-LAM description required for running a system-level rapid prototyping process.

This paper is organized as follows: related works dedicated to hardware resource modeling and IP-XACT integration are highlighted in Sect. 2. Section 3 introduces our framework for the co-design of MP2SoC embedded systems. Section 4 details our proposed S-LAM generator including the implemented meta-models and transformation rules. Finally, Sect. 5 gives some experimental results.

2 Related Work

In recent years, there has been an extensive interest in merging MDE-based frameworks and metadata IP reuse approaches. Initial efforts targeting to combine UML design entries with IP-XACT have been gaining traction [8–10]. These efforts aim to choose the adequate profile that covers the specification of complex hardware platforms on the one hand, and to implement the adequate mapping that generates the required IP-XACT description of the architecture on the other hand.

2.1 Using UML Profiles for HW Resource Modeling

UML is a general language but its extensibility, introduced with UML 2.0 via the notion of profiles, extends the language to domain-specific problems. More precisely, UML started to be adopted as a standard in the domain of real-time and embedded systems during the past years. Several profiling mechanisms aiming to use UML in SoC design and especially in hardware specification have been proposed including UML for SoC [11] and Omega-RT [12] profiles. With the ever increasing demand and complexity of embedded systems, a new profile has emerged. This standardized profile, named MARTE [6], is structured around two central concerns, modeling the characteristics of embedded systems and annotating the models to support the analysis of the system features. Defining accurate semantics for time and Hw/Sw resource modeling and supporting real-time and embedded systems co-design flows are the major goals of the MARTE profile. These two goals can be achieved using the MDE foundations when defining embedded system design flows. This explains the use of MARTE and MDE in the proposed co-design flow. In one hand, MDE facilitates automatic transformations from one abstraction level to a lower one, for simulation or implementation purposes. In the other hand, it promotes the integration of different tools thanks to transformation techniques. As a result, analysis tools, verification tools and modeling tools can be coupled in a single co-design flow.

2.2 Merging UML and IP-XACT in MDE-based Design Flows

Several works have shown the importance of integrating IP-XACT while taking advantage from MDE principles in their design flows. In [8] a MARTE-based methodology that exploits IP-XACT to specify and automatically generate Dynamic Partial Reconfiguration (DPR) SoC designs was proposed. MARTE models of the platform are parsed executing a chain of model transformations to obtain an IP-XACT description of the system that can be used in the Xilinx EDK (Embedded Design Kit) environment. In the COMPLEX framework [9], the IP-XACT description of the architecture can be automatically generated from the UML/MARTE model using the MARTE to IP-XACT (MARTIX) code generator [10]. Then, an executable model can be built from the IP-XACT platform description for functional validation and performance estimation. In another work [13], IP-XACT was used as input point in an MDE-based approach aiming to generate SystemC code. The authors propose a multi-level design flow that integrates extensions of the IP-XACT standard and different meta-models. Comparing these related works with our approach, we can observe that none of them uses IP-XACT for the high-level design space exploration of MP2SoC systems. Moreover, these works try to exploit the whole IP-XACT metadata targeting low-level simulations. On the contrary, our approach is based on a simplified sub-set of IP-XACT, named S-LAM, for the high-level analysis of MP2SoC.

3 A Co-Design Framework Integrating the S-LAM Generator

Our proposed approach, depicted in Fig. 1, is a complete EDA tool for the co-specification, design space exploration and code generation of MP2SoC systems that relies on Object Management Group (OMG) standards and MDE techniques. Being based on the Eclipse framework, front-end, transformation engine and back-end tools are grouped together in a fully-integrated flow.

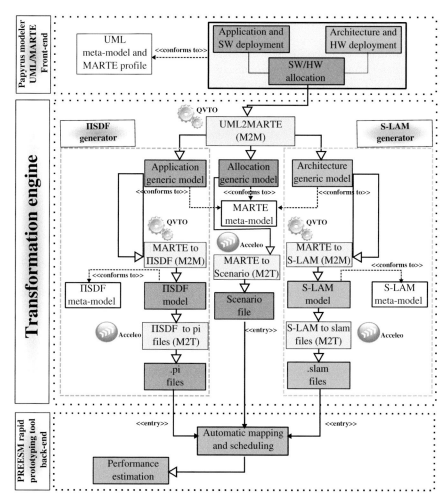

Fig. 1 The S-LAM generator in the context of the co-design flow

3.1 UML/MARTE Front-End

The proposed co-design flow uses UML/MARTE and the associated Papyrus tool [14] as modeling front-end. This high-level modeling front-end allows a user to graphically specify an embedded system conforming to the UML meta-model and the MARTE profile. Our methodology defines four sub-models to be specified and associated in a unified UML/MARTE based-model: application, architecture, allocation, and deployment sub-models.

3.1.1 Application Sub-model

Contains the structural specification of a given data-intensive application where computations are defined as a set of interconnected tasks inside a UML composite structure diagram. Application constraints and properties are defined in this sub-model including execution time value of each task using the «swSchedulableResource» stereotype from the MARTE *Software Resource Modeling* (SRM) sub-profile. The MARTE *Repetitive Structure Modeling* (RSM) sub-profile is used to model the parallel computations and the multidimensional data structures in the application. In addition, the *Generic Component Modeling* (GCM) sub-profile helps to define data flow ports and connectors.

3.1.2 Architecture Sub-model

Gathers a number of interconnected resources specifying the hardware components of an embedded system in a structural way. Therefore, the composite structure diagram is used to model the hierarchic structure of MP2SoC. Stereotypes from the MARTE *Hardware Resource Modeling* (HRM) sub-profile are exploited to indicate which kind of hardware component each UML element represents («HwProcessor» «HwMemory» «HwCommunicationResource» stereotypes). Properties of hardware processing resources, storage resources and communication resources are also specified using tagged values of these stereotypes. Multidimensional parallel resources of massively parallel MP2SoC architectures are specified using the *RSM* sub-profile. Ports and interconnections between hardware resources are annotated with stereotypes from the *GCM* sub-profile.

3.1.3 Allocation Sub-model

Defines the allocation constraints which associate tasks from the application sub-model with resources from the architecture sub-model. To allocate tasks to hardware components, the MARTE *alloc* sub-profile is used. In fact, UML dependencies between class instances of the application and the architecture are annotated with «allocate» or «distribute» stereotypes helping to map each task to a component or a repetition of a task to a group of components. The allocation is partial and defines only mapping constraints since the rapid prototyping tool automatically makes mapping decisions.

3.1.4 Deployment Sub-model

Describes the deployment of the software and the hardware components on IPs using the UML deployment diagram. The UML deployment mechanism and the MARTE profile lack aspects that allow the deployment of IPs on a component of the SoC. For this reason, our flow proposes an additional profile, the *Deployment* profile to facilitate deploying elementary components with IPs. The proposed profile facilitates both the high-level modeling of IPs and the automatic generation of the S-LAM system description. The «HwIP» stereotype, from the *Deployment* profile, models an IP deployed on a component of the architecture facilitating the generation of S-LAM descriptions. It gathers a set of attributes used to specify a component description in the S-LAM standard.

3.2 Transformation Engine

Three transformation engines were developed inside the transformation engine:

- **The π SDF generator:** produces π SDF graphs of the data-parallel application to facilitate the analysis of modern data-intensive applications running on MP2SoC architectures. The implementation of the π SDF generator is detailed in [15].
- **The S-LAM generator:** produces an S-LAM description of the architecture (cf. Section IV).
- **The MARTE to Scenario transformation:** produces a scenario file for the rapid prototyping framework. This scenario gathers systems constraints and properties aiming to guide the rapid prototyping process.

3.3 PREESM Tool Back-End

The generated π SDF graphs of the application, S-LAM description of the architecture and scenario file can be automatically analyzed and processed using the PREESM [7] rapid prototyping tool for automatic allocation, scheduling [16], system performance estimation [7] and finally code generation.

4 The S-LAM Generator

The implementation of a transformation flow in the MDE approach relies on the definition of ad-hoc meta-models for each abstraction level. For this reason, two meta-models are proposed in the context of the S-LAM generator: the MARTE meta-model and the S-LAM meta-model. In addition, model-to-model (M2M) and model-to-text (M2T) transformations were defined inside the transformation chains as depicted in Fig. 1. In our approach, M2M transformation rules are defined using the QVTO language [17] and M2T transformation rules are described using the Acceleo tool [18].

4.1 MARTE Meta-Model Relevant Parts Used in the S-LAM Generator

The input of each transformation chain in the proposed framework is a UML model compliant with the MARTE profile. Generating a MARTE model (conforming to the MARTE meta-model) from a profiled UML model (conforming to the UML meta-model) is a typical transformation in a UML/MARTE-based framework. The developed UML2MARTE transformation corresponds to a bridge connecting the specification of the system and the developed generators. This transformation is out of the scope of this paper. The open-source Ecore version of the MARTE meta-model provided with the source code of Papyrus and extended with the Deployment elements is used as the input of the S-LAM generator.

4.1.1 Conserving the Hierarchical Structure of MP2SoC with GCM Meta-Model

The *GCM* package from the MARTE profile defines a rich base of notations helping to annotate ports, interconnections, etc. However, supporting component-based models remains most important when focusing on moving up from specification purposes, where the MARTE profile is employed as a foundation, to successive transformations for DSE, where the MARTE meta-model is used as starting point. The *GCM* meta-model can preserve the hierarchical structure of a model without losing any detail since it represents an abstraction of the UML structured classes. A hierarchical component in MARTE is a *StructuredComponent* that encloses instances of other components, presented using the *AssemblyPart* element. Two assembly parts are connected via their ports (*FlowPort* element) using connectors. Connectors between two *AssemblyParts* are named *AssemblyConnectors*.

4.1.2 Capturing Repetitive Structures in the RSM Meta-Model

The *RSM* meta-model extends the basic concepts of the MARTE meta-model by providing meta-classes that capture shaped multiplicities and link topologies of intensive computation embedded systems. This meta-model proposes high-level meta-modeling mechanisms that express all the available parallelism of the hardware execution platform precisely and in a compact manner. These mechanisms are oriented toward two features: capturing the regularity of an MP2SoC system structure (composed of a repetition of structural elements) and denoting the topologies of links between hardware components of the system.

4.1.3 Capturing System Properties in the HRM Meta-Model

The *Hw_Logical* meta-model is the relevant part from the *HRM* meta-model used in the S-LAM generator as it gathers the set of hardware resources that are central to the MP2SoC platform definition. Properties of memories (size), communication networks (speedup) and processors can then be captured inside the meta-classes of this meta-model.

4.1.4 Capturing IP Properties in the Deployment Meta-Model

The Ecore version of the current MARTE meta-model was extended to enable its merging with the *Deployment* meta-model. Properties of each IP can be then deduced in the generated MARTE model from the «hwIP» stereotype and captured inside the *hwIP* meta-class.

4.2 The S-LAM Meta-Model

At high-levels of abstraction, a detailed description of each hardware resource is not necessary to succeed a rapid prototyping process. For this reason, the S-LAM meta-model does not use the entire IP-XACT meta-model, but it exploits a sub-set of concepts that capture the needed information for the exploration phase. This sub-set includes two meta-models: the *component* meta-model and the *design* meta-model.

4.2.1 The Component Meta-Model: Simplifying IP Description for DSE

A component, according to the IP-XACT standard, specifies a single hardware IP and details the required information for the integration of this IP including its interfaces and its internal structure. Assuming that a specification approach that ignores the implementation details of each component of the hardware architecture while detailing its primary properties makes the system-level exploration process faster and gives satisfactory solutions, the S-LAM *component* meta-model defines only three component types: operators, enablers and communication nodes. These components are efficient enough to specify a massively parallel embedded architecture that gathers processing elements (operators), local and shared memories (enablers) and regular and irregular communication networks (communication nodes).

4.2.2 The Design Meta-Model: Supporting Hierarchy and Composition

The S-LAM *design* meta-model, depicted in Fig. 2, describes a design as a set of component instances (*ComponentInstance* element), links (*Link* element), hierarchy

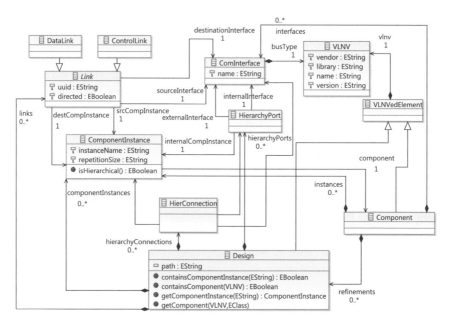

Fig. 2 S-LAM design meta-model

ports (*HierarchyPort* element) and hierarchy connections (*HierConnection* element). Both *Design* and *Component* elements are identified using their *VLNV* which specifies the vendor, the containing library, the element name, and the version number of a given IP. Each component instance in the design refers to the initial component description. These component instances can be connected using two types of connection elements: *Link* and *HierConnection*. While links are point-to-point connections between communication interfaces (*ComInterface* element) of the component instances, hierarchy connections connect sub-designs or components from different hierarchical levels using hierarchy ports. The original Ecore version of the S-LAM meta-model [7] was extended to allow the specification of a repetition of the same IP. The *repetitionShape* attribute was added to the *ComponentInstance* meta-class allowing to specify the repetition shape of a given component instance.

4.3 M2M Mapping Rules: From MARTE Model to S-LAM Model

The basic UML to MARTE and MARTE to S-LAM implemented QVTO mappings are sketched in Fig. 3.

UML meta-model	MARTE meta-model	S-LAM meta-model
Design		
UML::Class (with parts)	GCM:: StructuredComponent	slam::Design
UML::Property (part inside a hierachical class)	GCM:: AssemblyPart	slam:: ComponentInstance
UML::Port (port inside a hierachical class)	GCM::FlowPort (of a StructuredComponent)	slam:: HierarchyPort
UML::Connector (between two parts)	GCM::AssemblyConnector (between AssemblyParts)	slam::DataLink or slam::ControlLink
Component		
UML::Property (part inside a hierachical class)	GCM:: AssemblyPart	slam::Component
UML::Port (port inside a part)	GCM::FlowPort (of an AssemblyPart)	slam::ComInterface
UML::Property (stereotyped HwProcessor)	GCM::AssemblyPart (with a HwProcessor classifierTypeExtension	slam::Compoonent::Operator

Fig. 3 Mappings between UML, MARTE and S-LAM meta-models

4.3.1 Building the Hierarchical Structure of S-LAM

The S-LAM generator navigates the MARTE-compliant model and produces one S-LAM model. This model is produced if and only if the S-LAM generator finds at least one *StructuredComponent* in the MARTE model. Then, the hierarchical structure of the S-LAM model is created based on the *Design* meta-model. First, each *Structured-Component* is transformed into a *Design*. Each *AssemblyPart* within the *Structured-Component* becomes a *ComponentInstanse* inside the *Design* element. Moreover, if the shape of the *AssemblyPart* is superior to one, the *repetitionSize* attribute of the *ComponentInstance* will take the value of the *shape* element, indicating a repetition of a hardware component instance. Examining each *StructuredComponent*, the S-LAM generator looks for the *AssemblyConnectors* which associate *AssemblyParts*, and produces *DataLinks* or *ControlLinks* depending on the *AssemblyParts* type (HwProcessor, HwMemory, etc.). In addition, *AssemblyConnectors* linking an *AssemblyPart* with the *StructuredComponent* itself are transformed into *HierarchicalConnections*. For the production of *HierarchyPorts*, the generator explores the ports set of a given *StructuredComponent*, and transforms each *FlowPort* into a *HierarchyPort*.

4.3.2 Generating the Interface Set of Each Component Instance and Deducing its Type

For each *AssemblyPart* of the *StructuredComponent*, the S-LAM generator simultaneously produces a *ComponentInstance* and a *Component*. The implemented transformation automates the generation of the corresponding *ComInterfaces* of each *Component*. In fact, *FlowPorts* of each *AssemblyPart* are converted into *ComInterfaces* when mapping the corresponding *AssemblyPart* into *Component*. Furthermore, the generator is able to produce the right type of *Component* once it checks the *classifierTypeExtension* element attached to the *AssemblyPart*. In fact, if the *AssemblyPart* is not hierarchic, it will be transformed into an *Operator*, a *Mem* or a *ComNode* depending on its *classifierTypeExtension* (HW_Processor, HW_Memory, HW_CommunicationResource, HW_Bus). A hierarchical *AssemblyPart* is an instance of a *StructuredComponent* which was a hierarchical class stereotyped «HwResource» in the UML model. It is transformed into an *Operator* if it contains in its internal structure a processor.

4.4 M2T Mapping Rules: From S-LAM Model to S-LAM Files

Figure 4 shows the main Acceleo template which is the entry point of the M2T transformation. Given that this template requires an instance of the parameter *Design*, the transformation will navigate in the whole model to find all the available *Design* elements and generate one S-LAM file per Design. The produced files are named as the *Design* plus the ".slam" suffix, and encloses the «spirit:design» entry. Then, for each *ComponentInstance* element from the S-LAM model, the transformation will produce one component instance inside the «spirit:componentInstances» and «spirit:componentInstances» delimiters. At the same time, this transformation controls the *repetitionShape* value of each *ComponentInstance* in order to generate N (where N is the value defined by the *repetitionShape* attribute) component instances indicating the presence of a repetition of the same component instance in the design. The M2T transformation searches all the *DataLinks* and *ControlLinks* and produces a set of S-LAM interconnections. It also implements a similar navigation to figure out the list of hierarchical connections.

5 Case Study

To evaluate the benefits of our framework, we conduct a series of experiments on the M-JPEG encoder application. Originally developed for streaming multimedia application, the M-JPEG video compression format is now considerably exploited in video-capture devices where each video frame or video sequence is compressed separately as a JPEG image. Compared to the recently emerged video compression

```
[template public generateDesign(design : Design)]
[comment @main/]
[file (design.name.toString().concat('.slam'), false, 'UTF-8')]

<?xml version="1.0" encoding="UTF-8"?>
<spirit:design
xmlns:spirit="http://www.spiritconsortium.org/XMLSchema/SPIRIT/1.4">

[design.generateVLNV()/]

     <spirit:componentInstances>
        [for (compInstance:ComponentInstance | self.componentInstances)]
            [compInstance.generateComponentInstance()/]
        [/for]
     </spirit:componentInstances>

     <spirit:interconnections>
        [for (datalink:DataLink | self.links)]
            [datalink.generateDataLink()/]
        [/for]

        [for (controllink:ControlLink | self.links)]
            [controllink.generateControlLink()/]
        [/for]
     </spirit:interconnections>

     <spirit:hierConnections>
        [for (hierconn:HierConnection | self.hierarchyConnections)]
            [hierconn.generateHierConnection()/]
        [/for]
     <spirit:hierConnections/>

<spirit:vendorExtensions>
        [design.generatevendorExtensions()/]
</spirit:vendorExtensions>

</spirit:design>
[/file]
[/template]
```

Fig. 4 Acceleo main template

standards, M-JPEG describes a relatively simple encoding workflow. But, it is a typical streaming application that contains inherent task and data parallelism the fact that provides rich experimentation opportunities when running on MP2SoC architectures. Figure 5 shows the composite structure diagram of the application. The video sequence should first be partitioned into frames (M-JPEG_encoder class). Frames are split in blocks of 8*8 pixels and processed separately as JPEG images (Encode_Frame class). We performed experiments by simulating the M-JPEG on a stream of 100 and 200 frames of QCIF format (352 × 288 pixels). For this reason, multiplicities of tasks and ports expressed via the «Shaped» stereotype were varied.

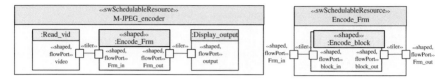

Fig. 5 UML/MARTE specification of the application

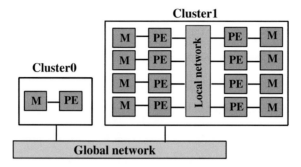

Fig. 6 MP2SoC architecture

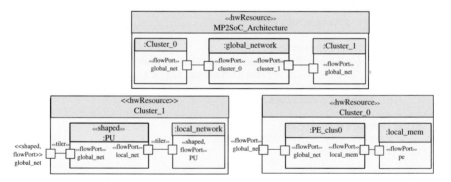

Fig. 7 UML/MARTE specification of the architecture

MP2SoC, as presented in Fig. 6, is composed of two clusters. While the first cluster contains one processing element (PE), the second cluster includes a variable number of processing elements. Processing elements inside the clusters are homogenous. Inside each cluster, each processing element is connected to its local memory and can communicate to other processors via a local network. The clusters can communicate via a global interconnection network. In order to model such complex system, a UML composite structure diagram is used as seen in Fig. 7. Each hierarchic hardware resource (MP2SoC system, clusters, processing units) is specified using a hierarchic class. For the rapid prototyping of the M-JPEG application, five configurations of MP2SoC were specified and generated varying the number of processing units (by changing the shape value of the PU class) containing 2, 4, 8, 24 and 32 processing

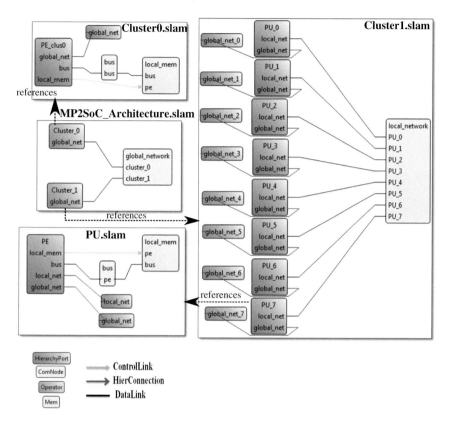

Fig. 8 Generated S-LAM files

units in Cluster1. Executing the S-LAM generator, four .slam files were created and visualized using the S-LAM editor as shown in Fig. 8. Each hierarchic class is transformed first into a Design element then into an .slam file. Class instances inside the hierarchic class are mapped into operators, memories or communication nodes. Hierarchic class instances that reference classes containing operators are transformed into operators that reference the .slam file that describes the internal structure of the classes themselves. Ports of the hierarchic classes becomes hierarchy ports. The «Shaped» annotation attached to the PU class and the port of the Cluster1 hierarchic class allows to produce eight hierarchy ports and link them with the eight operators with hierarchy connections in the MP2SoC configuration that contains eight processing units. π SDF files and the scenario file are also generated executing the two other transformation chains. The final step in the proposed approach is the rapid prototyping of the π SDF/S-LAM combination using PREESM. Figure 9 shows the average speedup of the application for two video sequence containing 100 and 200 frames respectively running on different MP2SoC configurations. We notice that for

Fig. 9 Speedup results

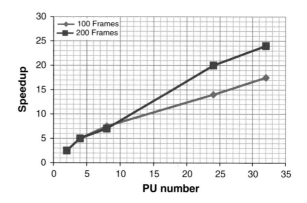

the video sequence containing 200 frames, increasing the PU number from 2 to 32 contributes for up to 10x M-JPEG encoder speedup. This observation justifies the use of MP2SoC architectures.

6 Conclusion

In this paper, the S-LAM generator, a tool able to generate S-LAM description of an MP2SoC architecture described in UML/MARTE under the proposed co-design flow specification methodology was presented. High-level models of the complex architecture are progressively refined enabling the production of a system-level description of the architecture for the design space exploration step, which is based on the PREESM framework. The S-LAM generator reduces the modeling effort as it starts from a co-specification of the whole MP2SoC system, including the application and the architecture parts, and captures needed information for the generation of IP-XACT compliant description of the architecture. Our next future work will be concentrated on the elaboration of a use case that takes as design entry a complex massively parallel application (An H.264 decoder for example) running on an MP2SoC architecture.

References

1. Engelmann, C., Lauer, F.: Facilitating co-design for extreme-scale systems through lightweight simulation. In IEEE International Conference on Cluster Computing Workshops and Posters, CLUSTER WORKSHOPS, 2010, pp. 1–8 September 2010
2. Lugato, D., Bruel, J-M., Ober, I.: Model-Driven Engineering for High Performance Computing Applications. In: S. Cakaj (ed.) Modeling Simulation and Optimization-Focus on Applications (2010)

3. Object Management Group. Unified Modeling Language specification, version 2.1. Available: http://www.omg.org/spec/UML/2.1.2/Infrastructure/PDF
4. Ecker, W., Müller, W., Dömer, R.: Hardware-Dependent Software, pp. 1–13. Springer, Netherlands (2009)
5. IEEE standard for IP-XACT, standard structure for packaging, integrating, and reusing IP within tools flows, IEEE Std 1685-2009, February 2010, pp. C1-360
6. Object Management Group. UML Profile for MARTE: Modeling and Analysis of Real-Time Embedded Systems, version 1.0. Available: http://www.omg.org/spec/MARTE/1.0/PDF/
7. Pelcat, M., Desnos, K., Heulot, J., Guy, C., Nezan, J.-F., Aridhi, S.: Preesm: A dataflow-based rapid prototyping framework for simplifying multicore DSP programming. In 6th European Embedded Design in Education and Research Conference. EDERC 2014, pp. 36–40 (2014)
8. Ochoa-Ruiz, G., Labbani, O., Bourennane, E.-B., et al.: A high-level methodology for automatically generating dynamic partially reconfigurable systems using IP-XACT and the UML MARTE profile. Des. Autom. Embed. Syst. **16**(3), 93–128 (2012)
9. Herrera, F., Posadas, H., Villar, E., Calvo, D.: Enhanced IP-XACT platform descriptions for automatic generation from UML/MARTE of fast performance models for DSE. In 15th Euromicro Conference on Digital System Design, DSD 2012, pp. 692–699, September 2012
10. Herrera, F., Villar, E.: A Framework for the Generation from UML/MARTE Models of IP-XACT HW Platform Descriptions for Multi-Level Performance Estimation. Proceedings of the Forum of Design and Specification Languages, FDL'2011, November 2011
11. Object Management Group. UML profile for System on a Chip, version 1.0. Available: http://www.omg.org/spec/SoCP/1.0/PDF/
12. Graf, S., Ober, I., Ober, I.: A real-time profile for UML. Int. J. Softw. Tools Technol. Trans. **8**(2), 113–127 (2006)
13. El Mrabti, A., Pétrot, F., Bouchhima, A.: Extending IP-XACT to support an MDE based approach for SoC design. In Design, Automation and Test in Europe Conference and Exhibition, DATE'09, pp. 586–589, April 2009
14. Papyrus, http://www.eclipse.org/papyrus/
15. Ammar, M., Baklouti, M., Pelcat, M., Desnos, K., Abid, M.: MARTE to π SDF transformation for data-intensive applications analysis. In Conference on Design and Architectures for Signal and Image Processing, DASIP, October 2014
16. Pelcat, M., Menuet, P., Aridhi, S., Nezan, J.F.: Scalable compile-time scheduler for multi-core architectures. In Proceedings of the Conference on Design, Automation and Test in Europe, DATE'09, pp. 1552–1555, April 2009
17. Guduric, P., Puder, A., Todtenhofer, R.: A comparison between relational and operational QVT mappings. In the 6th International Conference on Information Technology: New Generations, ITNG '09, pp.266–271, April 2009
18. Acceleo (2015) https://www.eclipse.org/acceleo/

Automatic Translation of OCL Meta-Level Constraints into Java Meta-Programs

Sahar Kallel, Chouki Tibermacine, Bastien Tramoni, Christophe Dony and Ahmed Hadj Kacem

Abstract In order to make explicit and tangible their design choices, software developers integrate, in their applications' models, constraints that their models and their implementations should satisfy. Various environments enable constraint checking during the modeling stage, but in most cases they do not generate code that would enable the checking of these constraints during the implementation stage. It turns out that this is possible in a number of cases. Environments that provide this functionality only offer it for functional constraints (related to the states of objects in applications) and not for architectural ones (related to the structure of applications). Considering this limitation, we describe in this paper a system that generates metaprograms starting from architecture constraints, written in OCL at the metamodel level, and associated to a specific UML model of an application. These metaprograms enable the checking of these constraints at runtime.

Keywords Software architecture · Architecture constraint · Object constraint language · Java reflect

---

S. Kallel (✉) · C. Tibermacine · B. Tramoni · C. Dony
Lirmm, Montpellier University, Montpellier, France
e-mail: sahar.kallel@lirmm.fr

C. Tibermacine
e-mail: chouki.tibermacine@lirmm.fr

B. Tramoni
e-mail: bastien.tramoni@lirmm.fr

C. Dony
e-mail: dony@lirmm.fr

A.H. Kacem
ReDCAD, Sfax University, Sfax, Tunisie
e-mail: ahmed.hadjkacem@fsegs.rnu.tn

213

R. Lee (ed.), *Software Engineering, Artificial Intelligence, Networking and Parallel/Distributed Computing 2015*, Studies in Computational Intelligence 612,
DOI 10.1007/978-3-319-23509-7_15

1 Introduction

Software architecture description is one of the main building blocks of an application's design [4]. It gives us an overview of the application organization that helps us to reason about certain properties, such as quality attributes. In this context, architecture description languages have been created to specify and verify such application architectures without worrying, at first, about the implementation of their functionality. The verification can be especially based on constraints that those languages associate to architecture descriptions. These constraints can be classified into two categories: functional and architectural.

Functional constraints check the state of the architecture's objects. For example, if we consider a UML model (an architecture description) containing a class Employee (a component in that architecture) which has an integer attribute age, a functional constraint presenting an invariant in this class could impose that the values of this attribute (slot of an object of that class) must be included in the interval [16–70] for all instances of this class. On the other side, architecture constraints analyze the structure of the application, and not objects states. For example, they define invariants (boolean conditions) imposed by the choice of a particular architectural style or pattern, like the layered architecture style [22]. All these constraints can be specified at design stage through a constraint language like the "Object Constraint Language" (OCL) [19], the OMG standard.

In the literature and practice of software engineering there exists a large number of architecture patterns [9, 11, 25] whose architecture constraints have been formalized. But unfortunately, currently architecture constraints can be checked only at design time on design artifacts; they are ignored in the implementation stage. Therefore, a part of the knowledge and the expertise in the implementation of a software project "evaporates". To guarantee that architecture pattern source code will not undergo changes during evolution in the implementation artifacts or at runtime, we need to find a way to check the associated architecture constraints at the implementation stage knowing that with OCL language (for example), we can not check the architecture constraints at this stage. We can opt to rewrite them entirely with languages used by the developers at that development stage. And this task of rewriting all these constraints is tedious, time consuming and error prone. Constraints on the two stages of development (design and implementation) are syntactically different but they are semantically equivalent (conditions on architecture descriptions that are present in the two stages). So why not generate the ones from the others, like code can be generated from UML models? Moreover, most of existing tools for model-to-text (code) generation do not consider the generation of code for constraints associated to models. For those which exist [1, 8, 18], they only translate functional constraints, and not architectural ones.

Considering these limitations, we propose a multi-steps process for translating OCL architecture constraints into Java code. The obtained Java code uses the introspection mechanism provided by the programming language (Java Reflect) to analyze the structure of the application. This choice is motivated by our willingness to use a standard mechanism without resorting to external libraries. Reflection (introspection) enables language users to analyze architectures and to examine the structure of their classes at runtime. In our work, the generated code is considered as a "metaprogram" since it uses the introspection mechanism of the programming language for implementing an architecture constraint.

The remaining of this paper is organized as follows. In the following section, we illustrate the input and the output of the proposed process to better understand the context of our work. These will serve as running examples throughout the paper. In Sect. 3, we present our general approach indicating the steps for generating constraints into Java metaprograms. Sections 4, 5 and 6 describe these steps in detail. Before concluding and presenting some perspectives, we discuss the related work in Sect. 7.

2 Illustrative Example

To introduce the context of our work, we present an example of an architecture constraint enabling the checking of the "MVC (Model-View-Controller) pattern" [21]. We assume that we have three stereotypes, allowing us to annotate the classes in an application which represent the view (View), the model (Model) and the controller (Controller). This constraint states that the classes stereotyped Model must not declare dependencies with the classes stereotyped View. This makes it possible, among other things, to have several views for the same model, and thus to uncouple these classes that play different roles in the pattern. In addition, the classes stereotyped Model must not have dependencies with the classes stereotyped Controller. This makes it possible to have several possible controllers for the model.

Using OCL and the UML metamodel (Fig. 1), we obtain the following constraint:

```
1  context Class inv :
2  self.package.profileApplication.appliedProfile
3  .ownedStereotype-> exists(s:Stereotype|s.name='Model')
4  implies
5  self.supplierDependency.client->forAll(t:Type |
6  not(t.oclAsType(Class).package.profileApplication
7  .appliedProfile.ownedStereotype->exists(s:Stereotype |
8  s.name='View' or s.name='Controller')))
```

Listing 1 MVC pattern constraint in OCL/UML

The first line in the Listing 1 declares the context of the constraint. It indicates that the constraint applies to each class of the application ; the meta-class Class is then the starting point for all navigations in the rest of the constraint. Lines 2 to 3 serve to collect the set of classes representing the model (having the stereotype `Model`) by using the navigation package.profileApplication.appliedProfile. ownedStereotype. UML metamodel allows us to get an applied stereotype only starting from the pack age that contains the modeling element (a class, in our case) and not from the element itself. The problem is resolved in some tools like RSA-IBM where the UML metamodel has been extended with an operation named *getAppliedStereotypes()*, which

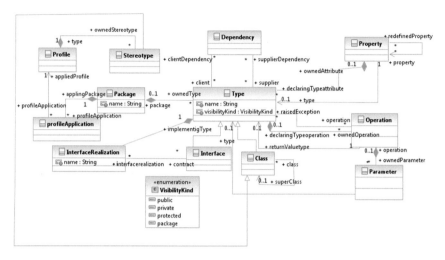

Fig. 1 An excerpt of UML metamodel

is inherited by the Class metaclass. In Line 5 we obtain the set of classes which have a direct dependency with the context of the constraint. The remaining of the Listing allows to iterate over the set of class instances and test if it contains classes stereotyped with `View` or `Controller`.

Our goal is to obtain automatically a metaprogram generated from an OCL/UML architecture constraint. The result would be expressed in Java as follows:

```
1  public boolean invariant(Class<?> aClass) {
2    if(aClass.isAnnotationPresent(Model.class)) {
3      Field[] fields = aClass.getDeclaredFields();
4      for(Field aField : fields){
5        Class<?> fieldType = aField.getType();
6        if(fieldType.isAnnotationPresent(View.class)
7        || fieldType.isAnnotationPresent(Controller.class))
8          return false;
9      }
10   }
11   return true;
12 }
```

Listing 2 MVC pattern constraint in Java

The method invariant(...) in Listing 2 accepts as a parameter an object of type Class, representing each of the classes of the application (the classes which compose the application business domain. This excludes classes of the libraries used by the application). Unfortunately, we cannot start navigation from the Package object representing the application package, because in java.reflect, this object does not enable to obtain references to the classes which are declared inside it. The Package object relates to a simple object containing information about the package(e.g. its name). We assume that the dependencies between classes in UML is translated as the declaration of at least one field in the first class having as a type the second class. In addition, we assume that the equivalent of stereotypes in UML are annotations in

Java. The method invariant (..) uses the Java reflect library by invoking, for example, getDeclaredFields() in Line 3 to collect fields, and isAnnotationPresent(..) in Lines 6 and 7 to check if a given type has been marked with a particular annotation.

3 General Approach

We propose a three-step process for generating executable Java code from architecture constraints. We note the presence of two metamodels the first one is the UML metamodel and the second is the Java metamodel that are presented in the following sections. Figure 2 depicts the process of metaprogram generation. If the OCL constraint needs a refinement, the first step consists in rewriting the OCL constraint in order to make it more accurate and concrete. For example, if the constraint has a navigation to Dependency metaclass (in UML metamodel) then we need to refine this constraint by specifying the different levels of dependencies. Else, the step of transforming OCL constraints from UML metamodel to Java metamodel is established in order to go forward in the process, to the Java code generation. These steps are detailed in the following sections. We did not perform a direct translation from OCL/UML to Java because this translation includes at the same time several transformations: shifting to a new metamodel, changing the syntax of constraints, etc. In fact, our approach requires first a mapping from abstractions of design level to abstractions of implementation level (mapping abstractions from UML metamodel to the Java metamodel) and subsequently a translation of the syntax.

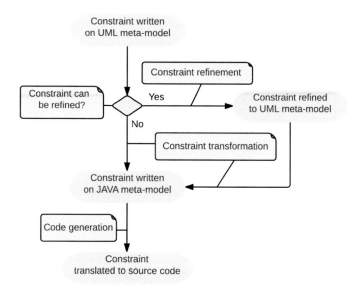

Fig. 2 Approach description

In the literature, there are many languages enabling the specification of architecture constraints (see [23] for a survey). The choice of OCL and UML is motivated by the fact that UML is the *de facto* standard modeling language, and that OCL is its original constraint language. Even if a recent study [20] pointed that UML is not widely used by developers, we all agree that it is a general-purpose modeling language known by a lot of developers. We have intuitively chosen to make constraints programmable in the implementation level in Java because it is a main-stream language in object-oriented programming, which provides introspection capabilities.

4 Constraint Refinement

The refinement mechanism is used whenever some abstractions in the UML metamodel do not have an equivalence in the JAVA language. For example, in the specification of the OCL constraint expressed on the UML metamodel, we have collected all types (Classes) which have dependencies with a specific type by using `supplierDependency.client`. This expression has not a direct equivalence in Java. As a result, we refine the constraint in the UML metamodel to express the different levels of dependencies.

Often, a dependency between two classes is translated as: (i) the declaration in the first class of at least one attribute having as type the second class, (ii) some parameters in operations of the first class, have as type the second class, or (iii) some operations of the first class, have as a return type the second class.

The previous constraint (Listing 1) is refined as follows:

```
 1  context Class inv :
 2    self.package.profileApplication.appliedProfile
 3    .ownedStereotype-> exists(s:Stereotype|s.name='Model')
 4  implies
 5    self.ownedAttribute.type->forAll(t:Type |
 6      not(t.oclAsType(Class).package.profileApplication
 7      .appliedProfile.ownedStereotype->exists(s:Stereotype |
 8      s.name='View' or s.name='Controller')))
 9  and
10    self.ownedOperation.returnValuetype->forAll(t:Type |
11      not(t.oclAsType(Class).package.profileApplication
12      .appliedProfile.ownedStereotype->exists(s:Stereotype |
13      s.name='View' or s.name='Controller')))
14  and
15    self.ownedOperation.ownedParameter.type->forAll(t:Type |
16      not(t.oclAsType(Class).package.profileApplication
17      .appliedProfile.ownedStereotype->exists(s:Stereotype |
18      s.name='View' or s.name='Controller')))
```

Listing 3 Refined MVC pattern constraint

Our constraint in Listing 3 (after refinement) is composed of three sub-constraints (Lines 5–8, Lines 10–13 and Lines 15–18). Each sub-constraint matches one level of the dependencies. In Line 5, the dependency is primarily verified on all attributes defined in classes. Note that oclAsType(Class) operation is used in this constraint to allow navigation between Type and Class through the specialization relation. In

Lines 10 and 15, the dependency is related to the types of operation parameters and their returned values.

The refinement of a constraint means a translation of this constraint from an abstract level to a concrete one. In contrast to the translation detailed in the following section, in this step, the translation is an endogenous transformation, since the constraints which are the source and the target of the transformation both navigate in the same (UML) metamodel.

5 Constraint Transformation

Before generating code, we transform in this step the OCL constraint specified on the UML metamodel into an OCL constraint specified on the Java metamodel. This simplifies the translation into Java code, since the mapping of abstractions from UML to Java is performed in this step. In order to perform constraint transformation we used a Java metamodel. Unfortunately, none of the metamodels found in the literature and practice satisfied our needs. We relied on Java Reflect library to create a new simplified Java metamodel. In fact, we can define our metamodel relying on Java specification but we deliberately chose Java Reflect because it gives us access to the meta-level of the language and also because it reflects exactly what we can do in the generated Java code. In this metamodel, we limited ourselves to the elements necessary for architecture constraint specification. Figure 3 depicts the Java metamodel that we have defined.[1]

The goal of constraint transformation is to replace in an architecture constraint the UML metamodel vocabulary by Java metamodel vocabulary. It had to establish a mapping between UML terms and Java terms that are classified in three categories: metaclasses, roles and navigations.

Table 1 presents for each UML metaclass, role and navigation its equivalent in Java.

We opted for the specification of these mappings in xml, and we have written an ad-hoc program for implementing the transformation instead of using an existing model transformation language like Acceleo [3], Kermeta [2] or ATL [16]. In fact, architecture constraints are not models. We might have generated models from constraints. But this process is complex to implement. It requires to transform the text of the constraint in models, to use a transformation language for transforming these models and then generate again the text of the new constraint from the new model. We opted for a simple solution that consists in exploiting an OCL compiler. It allows to generate an abstract syntax tree (AST) from the text of the constraint. This AST allows us to apply easily different transformations.

---

[1]We assume in this paper that the reader is familiar with UML and Java languages. This is the reason why the two metamodels are not detailed. They are depicted only for accompanying OCL constraints in order to see how navigations in the metamodels are established.

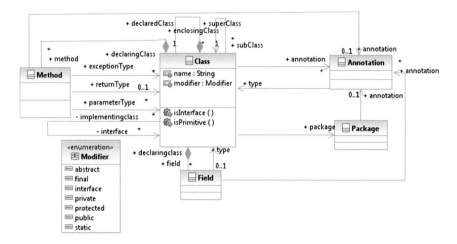

Fig. 3 Java metamodel

Table 1 Mapping UML-Java (Metaclass, Role, Navigation)

	UML	Java
Metaclass Role	**Class**	**Class**
	ownedAttribute	field
	ownedOperation	method
	superClass	superClass
	nestedType	declaringClass
	interfaceRealization	interface
	package	package
Navigation	package.profileApplication	
	.appliedProfile.ownedStereotype	annotation
Metaclass Role	**Property**	**Field**
	type	type
	declaringTypeattribute	declaringClass
Metaclass Role	**Operation**	**Method**
	returnValuetype	returnType
	declaringTypeoperation	declaringClass
	ownedParameter	parameterType
	raisedException	exceptionType
Metaclass	**Stereotype**	**Annotation**
Metaclass	**Package**	**Package**

We apply the table presented before (Table 1) on the generated AST in order to
obtain a constraint expressed in Java metamodel. For applying mappings, we start
by navigations, then the roles and finally the metaclasses. The following Listing 4
presents our constraint example after applying the transformation method:

```
1  context Class inv :
2    self.annotation–> exists(s:Annotation|s.name='Model')
3  implies
4    self.field.type->forAll(t:Type |
5    not(t.oclAsType(Class).annotation->exists(s:Annotation|
6    s.name='View' or s.name='Controller')))
7  and
8    self.method.returnType->forAll(t: Type |
9    not(t.oclAsType(Class).annotation->exists(s:Annotation|
10   s.name='View' or s.name='Controller')))
11 and
12   self.method.ParameterType->forAll(t: Type |
13   not(t.oclAsType(Class).annotation->exists(s:Annotation|
14   s.name='View' or s.name='Controller')))
```

Listing 4 MVC pattern constraint in OCL/Java

As indicated in Listing 4, we replace, among others, package.profileApplication.appliedProfile.ownedStereotype by annotation, ownedOperation by method, by respecting the mappings defined before.

The use of declarative mappings gives us the opportunity when the metamodels evolve to modify easily the changed elements. In addition, it allows us to offer a generic method which does not depend on particular metamodels.

6 Constraints Generation into Java Metaprograms

Code generation consists in translating the constraint expressed in Java metamodel into a Java metaprogram. To generate this code, we relied on the following steps. First, we generate the abstract syntax tree (AST) from the constraint expressed in Java metamodel. Then, when traverse this tree in a Depth-First Pre-Order way in order to generate progressively the java code by relying on rules presented below. It is worth mentioning that the first rule is applied only once in a constraint generation code. The other rules are applied along the analysis of the type of the AST nodes. In fact, if it is a role or navigation then we must apply Rule 2. If is a quantifier, the rule 3 is then applied and so on.

1. We must consider first that a constraint is represented by a Java method that returns a boolean, which takes as parameter an object of type the metaclass on which the constraint applies (its context). This method is located in a Java class and invokes if necessary other methods that are implemented during the code generation.
2. Each role and navigation in the Java metamodel will be transformed to its accessor method defined in Java. For example, if we navigate to Field, we apply getDeclaredFields(),[2] and if we would like to access to a method return type we call getReturnType().
3. Concerning the OCL quantifiers and the operations, we defined for each one a Java template. Examples are presented in Table 2. select(...) method presented in the last row of the table can be applied on different OCL collection types, like Set

---

[2]We use getDeclaredField() instead of getFields() to retrieve all attributes (private and public). For those we inherit, we must specify them in the OCL constraint using the role superClass.

Table 2 OCL Quantifiers and operations generation in Java

<table>
<tr><td rowspan="2">forAll</td><td>ocl</td><td>

```
forAll(ex:OclExpression): Boolean
```

</td></tr>
<tr><td>java</td><td>

```
private boolean forAll(Collection c) {
        for(Iterator i = c.iterator(); c.hasNext();) {
                if(!exInJava) return false;   }
        return true;
}
```

</td></tr>
<tr><td rowspan="2">exists</td><td>ocl</td><td>

```
exists(ex:OclExpression):Boolean
```

</td></tr>
<tr><td>java</td><td>

```
private boolean exists(Collection c) {
        for (Iterator i = c.iterator(); c.hasNext();) {
                ElementType e = (ElementType) i.next();
                if(exInJava) return true;   }
        return false;
}
```

</td></tr>
<tr><td rowspan="2">select</td><td>ocl</td><td>

```
select(ex:OclExpression):Sequence
```

</td></tr>
<tr><td>java</td><td>

```
List result = new ...();
private list select(Collection c) {
        for (Iterator i = c.iterator(); c.hasNext();){
                ElementType e = (ElementType) i.next();
                if (exInJava) {
                        result.add(e);   }
        }
return result;
}
```

</td></tr>
</table>

or Sequence. During the code generation , each OCL type will be replaced by its Java equivalent.

4. In each quantifier or operation, we traverse recursively the evaluated expression as a sub-constraint and we generate again the corresponding code: if we meet a role or navigation in Java metamodel, we re-apply rule 2. If the quantifier is nested, we re-apply rule 3, and so on.

5. In the case of a nested quantifier (two quantifiers for example are defined one inside the other), the second quantifier frequently needs to use the variables of the first one to define its expression. So, in this case, we store the variables of the first one (parameters of method that correspond to the first quantifier) in order to pass them among the parameters of the method corresponding to the second one.

6. Concerning the logic operators (and, not..), we defined also methods equivalent for each one. These methods are implemented in a class called LogicalOperator. If the constraint contains a logic operator, This class will be declared as a super class of the generated class that contains the invariant method.

7. The arithmetic operations ($>$, $<$, $=$, ...) and the types (Integer, Real, String, ...) are the same in the generated metaprogram.

In order to better explain the code generation process, Table 3 presents an example of a metaprogram which is generated from our MVC constraint presented in Sect. 2.

Table 3 Example of MVC constraint Code generation

Constraint	Java Metaprogram
| ```
context Class
  inv :
``` | ```
/*Rule 1*/

public class Constraint{
Boolean invariant(Class aClass){
  //To be completed }
}
``` |
| ```
self.annotation
``` | ```
/*Rule 2*/

Annotation [] annotations=
  aClass.getAnnotations();
``` |
| ```
->
exists
(a:Annotation|
 a.name='Model')
``` | ```
/*Rule 3*/

resultexists1 =
    exists1(annotations);
```<br>```
/* Rule 4*/

private Boolean exists1(Annotation[]
                         annotations)
{
 for(Annotation annota: annotations){
  Class a = annota.annotationType();
  if(a.equals(Model.class)){
      return true;   }
    return false;
}
``` |
| ```
self.method
``` | ```
/*Rule 2*/

Method[] methods=
  cl.getDeclaredMethods();
``` |
| ```
->
forAll(m:method|
not(m.returnType
   .annotation
->
exists
(a:Annotation|
a.name='View'))
)
``` | ```
/*Rule 3*/

Boolean resultforAll1=
  forAll1(methods);
```<br>```
/*Rule 4 and Rule 6*/

private Boolean forAll1(Method[]
                          methods)
{
 for(Method m : methods){
  Type type = m.getReturnType();
  Annotation[] annotations =
       type.getAnnotations();
  resultexist2 =
      not(exists2(annotations));
  if(!resultexist2) return false; }
    return true;
}
``` |

For simplicity reasons, we consider for the dependency between two classes that the first class has at least one method return type having as type the second class.

We have presented in Table 3, for each part of constraint, its equivalent Java code by respecting the rules that was explained previously. The generated code uses the introspection of Java in order to examine the application structure at runtime (`getAnnotations()`, `getMethods()`). This code should be called before and after each method and affectation implemented in the application.

It is worth noting that this code is syntactically different from the optimal code presented at the beginning of the paper (see Listing 2) but they are semantically equivalent. It is evident that the automatic translation does not allow to obtain a code having an optimal complexity. However, it is a valuable tool for developers who will rather focus on implementing the business logic of their application.

7 Related Work

In this section we present works related to OCL constraint transformation and OCL code generation. Hassam et al. [13] proposed a method for transforming OCL constraints during UML model refactoring using model transformations. Their approach uses first an annotation method for marking the initial UML model, in order to obtain an annotated target model. Then, a mapping table is created from these two annotations in order to transform OCL constraints of the initial model into OCL constraints of the target one. Their solution of constraint transformations is difficult to establish and it needs some knowledge about model transformation languages and tools. In our work, constraint transformation is simple. It is performed in an ad-hoc way without using additional modeling and transformation languages. In [10], the authors propose an approach to generate (instantiate) models from metamodels taking into account OCL constraints, using CSP (Constraint Satisfaction Problem). They defined some mathematical rules to transform models and constraints associated to them. Cabot et al. [7] worked also on UML/OCL transformation into CSP in order to check quality properties of models. These approaches are similar to our transformation process because they use an OCL compiler (DresdenOCL [8]) to transform constraints. But in our approach, we consider source code generation from these constraints, in order to make them executable with application's code. In contrast to CSP, this does not require an external tool for the interpretation of constraints.

In the practice of model-driven engineering, there exist several tools like Eclipse OCL [1], Octopus [18], and DresdenOCL [8, 14, 17] which aim to translate OCL constraints in Java source code. They however transform constraints which are functional and not architectural. These tools translate this kind of constraints into object-oriented programs which do not use the introspection mechanism. The generated code by Dresden OCL is difficult to understand. Indeed, it is true that Dresden OCL is the first tool implemented in this domain, but it extensively uses a vocabulary proposed only by its APIs. This code is normally intended to developers who master, and will continue to use, Dresden OCL, contrary to our work, where code is intended to be used by any Java developer. Besides, with these tools, we need to create beforehand the classes of the model before generating constraints. Other works like Briand et al. in [6] and Hamie et al. in [12] proposed a tool to transform functional (and not architectural) constraints respectively into Java using aspect-oriented programming and JML contracts.

8 Conclusion

It has been demonstrated that architecture constraints bring a valuable help for preserving architecture styles, patterns or general design principles in a given application after having evolved its architecture description [24]. These architecture constraints are checked at design time. But what if the architecture evolves in the implementation artifacts (the application's programs)? Or, what if the architecture evolves at runtime (through dynamic adaptation, for example)? To be able to check these constraints in that development stage and at runtime, architecture constraints should be translated into an appropriate format: meta-programs.

We have presented in this paper a process for generating Java code starting from OCL architecture constraint specifications expressed in the UML metamodel. This Java code uses the introspection mechanism provided by the programming language. Our process is composed of three steps. The first optional one consists in refining the constraints. The second step allows to transform them into OCL constraints expressed in Java metamodel. The last step generates Java source code relying on specific code generation rules. The reflection (instrospection) mechanism used in our approach is a standard mechanism in Java. Otherwise, we can use static analysis libraries like JDT [15] or ByteCode libraries like BCEL [5] but our goal was to use what is standard in Java and not resort to external libraries. In addition, with reflection, architecture constraints can be checked at runtime (by invoking the invariant method in all the methods of the application where the architecture is changed: new objects are created, references to objects are assigned to fields, etc.).

In our proposal, OCL coverage is not complete. We have implemented a prototype called *MOJaRT: Meta-OCL to JAva Reflect Translator*. It is available for download here: https://github.com/saharkallel/mojart.git/. which does not take into consideration some OCL constructions, like some collection operations (union, for example). But this does not have any impact on the work proposed in this paper.

As a future work, we plan to generalize the proposed approach, by specifying architecture constraints in a language-independent way: using predicates on graphs and operations on them and then making automatic transformations towards a particular object-oriented programming language.

References

1. Eclipse ocl. http://www.eclipse.org/modeling/mdt/?project=ocl
2. Kermeta. http://www.kermeta.org
3. Acceleo: Implementation of MOF to text language. http://www.omg.org/news/meetings/tc/mn/specialevents/ecl/Juliot-Acceleo.pdf
4. Bass, L., Clements, P., Kazman, R.: Software architecture in practice. Addison-Wesley, Boston (2012)
5. BCEL: The byte code engineering library. http://commons.apache.org/proper/commons-bcel/
6. Briand, L.C., Dzidek, W., Labiche, Y.: Using aspect-oriented programming to instrument OCL contracts in Java. Technical Report, Carlton University, Canada (2004)
7. Cabot, J., Clarisó, R., Riera, D.: Umltocsp: a tool for the formal verification of UML/OCL models using constraint programming. In: Proceedings of the 22nd IEEE/ACM International Conference on Automated Software Engineering, pp. 547–548. ACM (2007)
8. Demuth, B.: The dresden OCL toolkit and its role in information systems development. In: Proceedings of the 13th International Conference on Information Systems Development (ISD2004) (2004)
9. Erl, T.: SOA Design Patterns. Pearson Education, London (2008)
10. Ferdjoukh, A., Baert, A.E., Chateau, A., Coletta, R., Nebut, C.: A CSP approach for metamodel instantiation. In: IEEE Internationnal Conference on Tools with Artificial Intelligence, ICTAI 2013, pp. 1044–1051 (2013)
11. Gamma, E., Helm, R., Johnson, R., Vlissides, J.: Design patterns: Elements of Reusable Object-Oriented Software. Addison Wesley (1994)
12. Hamie, A.: Pattern-based mapping of OCL specifications to JML contracts. In: 2014 2nd International Conference on Model-Driven Engineering and Software Development (MODELSWARD), pp. 193–200. IEEE (2014)
13. Hassam, K., Sadou, S., Fleurquin, R., et al.: Adapting OCL constraints after a refactoring of their model using an MDE process. In: Belgian-Netherlands software Evolution seminar (BENEVOL 2010), pp. 16–27 (2010)
14. Hussmann, H., Demuth, B., Finger, F.: Modular architecture for a toolset supporting OCL. In: l UML2000The Unified Modeling Language, pp. 278–293. Springer, Berlin (2000)
15. JDT: Java development tools. http://www.eclipse.org/jdt/
16. Jouault, F., Kurtev, I.: Transforming models with ATL. In: Satellite Events at the MoDELS 2005 Conference, pp. 128–138. Springer, Berlin (2006)
17. LCI: Object constraint language environnement. http://lci.cs.ubbcluj.ro/ocle/
18. Octopus: OCL tool for precise UML specifications. http://octopus.sourceforge.net
19. OMG: Object constraint language, version 2.3.1, document formal/2012-01-01. http://www.omg.org/spec/OCL/2.3.1/PDF/
20. Petre, M.: UML in practice. In: Proceedings of the 35th International Conference on Software Engineering (ICSE 2013), pp. 722–731. IEEE Press (2013)
21. Reenskaug, T.: Thing-model-view editor an example from a planning system, xerox parc technical note (1979)
22. Shaw, M., Garlan, D.: Software Architecture: Perspectives on an Emerging Discipline. Prentice Hall, Upper Saddle River (1996)
23. Tibermacine, C.: Software Architecture 2. Software Architecture: Architecture Constraints. Wiley, New York (2014)
24. Tibermacine, C., Fleurquin, R., Sadou, S.: On-demand quality-oriented assistance in component-based software evolution. In: Proceedings of the 9th ACM SIGSOFT International Symposium on Component-Based Software Engineering (CBSE'06), pp. 294–309. Springer LNCS, Vasteras, Sweden (2006)
25. Zdun, U., Avgeriou, P.: A catalog of architectural primitives for modeling architectural patterns. Inf. Softw. Technol. **50**(9), 1003–1034 (2008)

Towards a Formal Model for Dynamic Networks Through Refinement and Evolving Graphs

Faten Fakhfakh, Mohamed Tounsi, Ahmed Hadj Kacem and Mohamed Mosbah

Abstract Due to the highly dynamic behavior and the time complexity in Mobile Ad-hoc NEtworks (MANETs), modeling distributed algorithms and looking at their assumptions represent a challenging research task. Also, proving the correctness of these algorithms for dynamic networks is a topic of intensive research. In fact, the solutions which have been proposed to express and prove the correctness of distributed algorithms are usually done manually. In addition, all these solutions lack a consensus about their development and their proof. The main contribution of this paper is to propose a general and formal model for dynamic networks based on evolving graphs and Event-B formal method. In fact, evolving graphs is a powerful tool to express fine-grained properties. This model allows to handle topological events and to characterize the concept of time with some particularities. We implement it with Event-B, based on refinement technique. To illustrate the proposed model, we investigate an example of a distributed algorithm encoded by local computations models.

F. Fakhfakh (✉) · M. Tounsi · A.H. Kacem
ReDCAD Laboratory, FSEGS, University of Sfax, B.P. 1088, 3018 Sfax, Tunisia
e-mail: faten.fakhfakh@redcad.org

M. Tounsi
e-mail: mohamed.tounsi@fsegs.rnu.tn

A.H. Kacem
e-mail: ahmed.hadjkacem@fsegs.rnu.tn

M. Mosbah
LaBRI Laboratory, University of Bordeaux, CNRS UMR 5800, 33405 Talence, France
e-mail: mohamed.mosbah@labri.fr

© Springer International Publishing Switzerland 2016   227
R. Lee (ed.), *Software Engineering, Artificial Intelligence, Networking and Parallel/Distributed Computing 2015*, Studies in Computational Intelligence 612,
DOI 10.1007/978-3-319-23509-7_16

1 Introduction

1.1 Overview

In recent years, wireless communication networks have witnessed rapid advances in the computing industry and are widely available in our everyday life. A MANET [13] is a form of wireless networks. It is composed of mobile computing devices, called nodes, such as laptops, smartphones, etc. These nodes are dynamically connected in an arbitrary manner, without the support of any fixed infrastructure or centralized administration. MANETs cover a large range of applications like military operations, emergency relief, wireless sensor networks, etc.

Due to node mobility, disconnections and failures that can be produced, MANETs are extremely dynamic and the connections between nodes vary in time. One well-known challenge in these networks is modeling such dynamics and creating a reference model on which results could be compared and reproduced. In this context, a MANET can be naturally represented as a dynamic graph whose nodes are mobile devices and the edges are instantaneous wireless links between the nodes. The evolving graph formalism has been proposed by A. Ferreira [8] as a combinatorial model for dynamic networks. In this model, a dynamic graph can be decomposed as a discrete sequence of static graphs. Each static graph is a snapshot of the dynamic network at a given time.

In a dynamic graph, the communication between nodes can be ensured by a distributed algorithm [14]. The latter is designed to run on interconnected autonomous computing entities for achieving a common task. In order to encode distributed algorithms, we use local computations model and particularly graph relabelling systems [12]. In this context, a node can realize a computation step if there is a specific rule that describes the corresponding label modifications. The rule can be applied if it is consistent with the states of the node and its neighbours.

To specify the abstraction provided by local computation, we use a formal method. In fact, formal methods provide a real help for expressing correctness with respect to safety properties in the design of distributed algorithms. Particularly, the *correct-by-construction* approach [11] provides a way to prove algorithms. It can be supported by a progressive and incremental process controlled by the refinement [3] of models for distributed algorithms. This process allows to simplify the proofs and to validate the integration of requirements. The Event-B modeling language [1] can support this methodological proposal suggesting proof-based guidelines. It is supported by a tool called "RODIN" [2] which provides an environment for developing *correct-by-construction* models for software-based systems.

1.2 Contribution

In this paper, we propose a reusable model for dynamic networks by combining the evolving graphs formalism and the refinement approach of the Event-B method. More precisely, we develop a general and formal solution which defines the different topological events in a dynamic network and the resulting changes in the evolving graph. In addition, we focus in this model on the concept of time to present the situations which need a time evolution. Moreover, the proposed model gives primitives to analyze an evolving graph. Our model takes into consideration only the variation of edges in the network. It can be extended in the future to address the movements of nodes.

Formally, we propose a refinement strategy that allows to enrich a model in a step by step fashion. The refinement is the foundation of the *correct-by-construction* which is a well suited approach to prove algorithms. The main objective of our model is to enable reuse during the development. In fact, different components of the model can be refined and reused to specify distributed algorithms in dynamic networks. Hence, we can save effort on proving correct algorithms.

To illustrate our model, we present an example of a counting algorithm encoded by local computations models. The main goal of this example is to demonstrate how we can use and refine the proposed model.

1.3 Related Work

Evolving graphs are an effective and powerful formalism which helps to capture the dynamic behavior of MANETs. That's why, it has drawn the attention of the research community in the last few years. Several research works have been based on this formalism to deal with network dynamics.

In [6], A. Casteigts proposed an analysis framework for distributed algorithms on dynamic networks. The proposed framework provides general formalisms and methods for studying the main properties of the distributed algorithms in dynamic networks. It allows to characterize the necessary and/or sufficient connectivity conditions required for the success of a distributed algorithm in a dynamic network. It is based on the combination of the evolving graphs and graph relabellings [12]. This framework is illustrated by the analysis of three simple algorithms (propagation algorithm, centralized counting and decentralized counting) whose necessary and sufficient conditions were derived into a sketch of classification of dynamic networks.

Furthermore, P. Floriano et al. [9] presented a study of necessary and sufficient conditions, in dynamic networks, for two distributed problems which are mutual exclusion and K-mutual exclusion. To do this, they exploit the framework proposed by A. Casteigts [6].

The author [10] provided a sufficient condition for the decentralized counting algorithm suggested by A. Casteigts [6]. In fact, he shows that a complete underlying graph was sufficient for the decentralized counting algorithm to succeed. Moreover, he introduces the concept of tight conditions, to strengthen the guarantees offered by necessary and sufficient conditions. Then, he demonstrates the tightness of the sufficient condition provided for the decentralized counting algorithm.

In addition, M. Barjon et al. [4] proposed an algorithm which maintains a forest of spanning trees in dynamic networks. The proposed algorithm aims to maintain exactly one token (root) per tree. It is based on three operations on tokens: circulation, merging and regeneration. To do this, a computation step takes as input the state of a pair of nodes and modifies these states according to some rules.

Throughout the related works outlined above, we note a lack of consensus about the development and proof of distributed algorithms in dynamic networks. Moreover, the proofs which have been presented are done manually. In addition, most of the distributed algorithms which have been investigated, in dynamic networks, are simple.

1.4 Organization of the Paper

The remainder of this paper is organized as follows: In Sect. 2, we present basic concepts of the evolving graph and the Event-B formal method. Section 3 introduces our proposed model for dynamic networks based on the evolving graphs formalism. In Sect. 4, we present the formal development of the proposed model. Section 5 applies our model to develop an example of a distributed algorithm. Finally, the last section concludes and outlines areas for our future research.

2 Preliminaries

2.1 Evolving Graphs

The formalism of evolving graphs has been proposed as a combinatorial model for dynamic networks. In this model, the evolution of the network topology is simply recorded as a sequence of static graphs. As an example, we consider the four snapshots taken at different time intervals of a MANET, as shown in Fig. 1. Each static graph is a snapshot of the dynamic network at a given time. This view is precisely adopted by A. Ferreira [8].

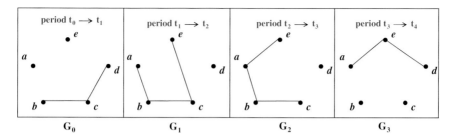

Fig. 1 Sucessive snapshots of a MANET evolution over time

Fig. 2 The evolving graph corresponding to the MANET in Fig. 1

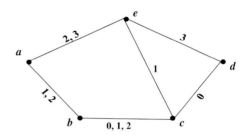

Formally, an evolving graph g is a triplet (G, S_G, $S_\mathbb{T}$), where:

- $S_\mathbb{T} = t_0, t_1, \ldots, t_n$ is an ordered sequence of dates used to capture the static graphs. These dates correspond to every time step in a discrete-time system ($\mathbb{T} \subseteq \mathbb{N}$). Except for t_0 and t_n, each t_i corresponds to one or more topological events that modifies the network. Each edge is labeled with the dates of its presence.
- $S_G = G_0, G_1, \ldots, G_{n-1}$ is the sequence of undirected static graphs. Each G_i represents the network topology during the period $[t_i, t_{i+1}[$ in the evolving graph g.
- G represents the union of all G_i in S_G, called the underlying graph of g (see Fig. 2). The edges are labeled with the date of their presence. For example, the presence of the edge *"ae"* in Fig. 1 at the dates "2" and "3" is represented in Fig. 2 by an edge *"ae"* labeled "2, 3".

We will use the simple notations "V" and "E" to denote respectively the sets of nodes and edges of the underlying graph "G".

2.2 Event-B Overview

The Event-B modeling language [1] defines mathematical structures into contexts and the formal model of the system into machines. The modeling process starts by identifying the domain of the problem expressed by means of context. This latter is characterized by a list of sets, list of constants, list of axioms and theorems that can be derived from the axioms of the context. An Event-B machine describes a reactive system by a set of invariants properties and a finite list of events modifying state variables. A machine "M" may see a context "C", this means that all carrier sets and constants defined in "C" can be used in "M".

The key tool behind the Event-B method is the *refinement* [3]. The refinement of a specification allows to enrich it in a *step-by-step* fashion. It is the foundation of the *correct-by-construction* approach. It provides a way to strengthen invariants and add details to a model. It is also used to transform an abstract model into a more concrete version by modifying the state definition. This is done by extending the list of state variables, by refining each abstract event into a corresponding concrete version and by adding new events.

An Event-B specification is considered as correct only if each machine, as well as the process of refinement, are proved by adequate theorems named Proof Obligations (PO). The management of proof obligations is a technical task supported by RODIN tool [2], which provides an environment for developing *correct-by-construction* models for software based systems.

3 The Proposed Model

Based on evolving graphs, we propose in this section a formal and general model for dynamic networks. The proposed model defines the different topological changes in a dynamic network allowing to specify a distributed algorithm, the manner of time evolution and the primitives to analyze the evolving graph. The main objective of this model is to be reused or instantiated to specify distributed algorithms in dynamic network.

As mentioned earlier, the formalism of evolving graphs allows to represent the changing connectivity of a dynamic network as a sequence of static graphs.

Let $g = (G, S_G, S_T)$ be an evolving graph. Every static graph, $G_i \in S_G$, corresponds to the network topology during the interval of time $[t_i, t_{i+1}[$ where "t_i" represents the date when one or several topological events occur in the system. In our work, the time evolution from a date "t_i" to a date "t_{i+1}" is performed after one or many topological events. We can distinguish two situations of these events:

- Adding edge:
  Pre-condition: appearance of a new edge in the network at the current date "t".
  Post-condition: addition of the new edge labeled "t".
- Removing edge:
  Pre-condition: presence of an edge in the network at the date "t-1" and its disappearance at the current date "t".
  Post-condition: no change takes place in the evolving graph.

There is another event that requires changing the evolving graph without affecting the time evolution. We call this event "Maintaining edge". We talk about this situation if an edge is present at the date "*t-1*" and it undergoes no change at the date "*t*". In this case, we add the date "*t*" to the label of the concerned edge.

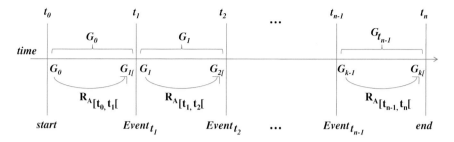

Fig. 3 Combination of graph relabellings and evolving graphs

In our proposed model, we take into account some hypotheses. On the one hand, we will consider only the variation of edges in the network. In contrast, we don't consider the appearance or disappearance of nodes. On the other hand, a distributed algorithm in the local computation model is simply given by a set of relabelling rules. So, we suppose that such distributed algorithm can apply its rewriting rule(s) to every edge before the final date "t_n". In order to analyze distributed algorithms on dynamic networks, we combine the evolving graphs and the formalism of graph relabellings [12]. As illustrated in Fig. 3, each static graph G_i in S_G covers the time interval $[t_i, t_{i+1}[$. We denote by "$Event_{t_i}$" the one or more topological events occurring at the time "t_i". Between two consecutive topological events, any number of relabellings may take place. For a given algorithm A and two consecutive dates $t_i, t_{i+1} \in S_{\mathbb{T}}$, we denote by:

- $G_{i[}$ the labeled graph representing the network state just before "$Event_{t_i}$";
- G_i the labeled graph representing the state of the network just after the topological events of the date "t_i";
- $R_{A_{[t_i, t_{i+1}[}}$ one of the possible relabelling sequence induced by the algorithm A on the graph G_i during the period $[t_i, t_{i+1}[$.

Then, we have $Event_{t_i}(G_{i[}) = G_i$ and $R_{A_{[t_i, t_{i+1}[}}(G_i) = G_{i+1[}$.

4 Formal Development

We remember that the specification of our proposed model is performed with Event-B method and done with RODIN platform. In our work, the development strategy of our model is composed of one context "c" and two machines "$M0$" and "$M1$". We begin by presenting the context which describes static properties of the network. After that, we detail the specification of the two machines. In fact, we start with a very abstract model and then we add details, to obtain a correct and concrete model.

4.1 Formal Development of the Context "c"

The context "c" describes the static properties of the network. Formally, a graph namely "g" is modeled by a set of nodes called "V". However, we have supposed in our work that the dynamic graph is composed of stable nodes and variable edges. For this reason, we define "V" in the context as an abstract set. Moreover, we add "tn" as a constant which represents the final system date. By means of the "axm1", we state that "tn" is an integer different to the start date of the system. Furthermore, we add "axm2" to indicate that the number of nodes in the network is finite. The axioms specification of the context "c" is done as follows:

$$axm1 : tn \in \mathbb{N} \setminus \{0\}$$
$$axm2 : finite(V)$$

4.2 Formal Development of the First Level: Machine M0

In this level, a network can be formally modeled as a connected, undirected and simple graph "g" where nodes denote processors and edges denote direct communication links (see inv1). A graph is undirected if there is no distinction between two nodes associated with each edge (see inv2). A simple graph means that it does not have more than one edge between any two nodes and no edge starts and ends at the same node (see inv3). The invariants specification of M0 is done as follows:

$$inv1 : g \subseteq V \times V$$
$$inv2 : g = g^{-1}$$
$$inv3 : (V \triangleleft id) \cap g = \emptyset$$

In the first level, we can notice the appearance of new edges and the maintain of the existing ones from a graph G_i to the following graph G_{i+1}. In fact, the basic idea of the evolving graph is the superposition of graphs one another. As a consequence, we do not consider the concept of time and the removing of edges. Nevertheless, it is necessary to have another level which refines the first. In Fig. 4, we present an example of evolving-graphs sequence which we can see in the first level. This sequence corresponds to the network topology taken in Fig. 1. Formally, we introduce two events "Adding_Edge" and "Maintaining_Edge".

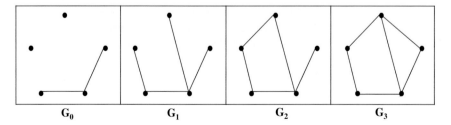

G_0 G_1 G_2 G_3

Fig. 4 Example of evolving-graphs sequence in M0

- **Adding_Edge:** In this event, if an edge does not belong to the graph "g" (grd1, grd2 and grd3), then we can add it to "g" (act1). To respect the invariant "inv2", we add both "$x \mapsto y$" and "$y \mapsto x$". We provide below the specification of this event.
- **Maintaining_Edge:** Based on the evolving graphs, if an edge exists in a graph G_i then it still exists in the following graph G_{i+1}. Formally, in the guard component, we define "grd1" to verify the existence of an edge "e" in the graph "g". This event with no action is considered to have the action *skip*.

```
Adding_Edge
any   x, y
where
   grd1 : x ↦ y ∈ V × V
   grd2 : x ↦ y ∉ g
   grd3 : x ≠ y
then
   act1 : g := g ∪ {x ↦ y, y ↦ x}
end
```

```
Maintaining_Edge
any   e
where
   grd1 : e ∈ g
then
   act1 : skip
end
```

4.3 Formal Development of the Second Level: Machine M1

The second machine, called *M1*, refines the previous one. In fact, we keep the variables, invariants and events of the machine *M0* and we add details to transform an abstract model into a more concrete version. If an invariant refers to both the abstract and concrete model, we call it a *"gluing invariant"* (inv3, inv4). The *gluing invariants* are used to relate the states between the concrete and abstract machines. In this level, we introduce the concept of time to distinguish the situations of appearance, disappearance and maintain of an edge in the network. Indeed, each edge has a label that indicates the dates when it is present in the network.

- The appearance of a new edge in the network at the date "t" requires the addition of this edge labeled "t";
- The disappearance of an edge does not change anything in the evolving graph;
- The presence of an existing edge at the current date "t" requires adding the date "t" to the label of the concerned edge.

To illustrate these details, we present an example of evolving-graphs sequence in Fig. 5 which refines the first level.

In order to specify these events, we refine the *"Adding_Edge"* event of M0 by another event *"Adding_Edge"* which add more details. Also, we refine the *"Maintaining_Edge"* event to obtain two events *"Maintaining_Edge"* and *"Removing_Edge"*. Moreover, we introduce a new event called *"Incrementing_Time"*, to ensure the incrementation of time when one or many topological changes occur in the network.

Formally, we specify the machine *M1* by adding two variables "t" and "LE". The variable "t" represents the current time and "LE" is a function that assigns a label to each edge. The specification of this function takes the following form:

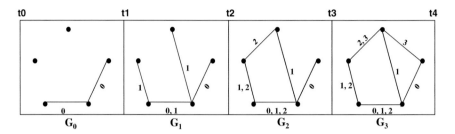

Fig. 5 Example of evolving-graphs sequence in *M1*

$LE \in g \rightarrow \mathbb{P}(N)$. The addition of these two variables involves the addition of new properties in the invariant component (inv1, inv2, inv3 and inv4).

$$
\begin{array}{l}
inv1 : LE \in g \rightarrow \mathbb{P}(\mathbb{N}) \\
inv2 : t \in \mathbb{N} \land t \geq 0 \land t \leq tn \\
inv3 : \forall e \cdot e \in g \implies (\exists t1 \cdot t1 \in \mathbb{N} \land t1 \geq 0 \land t1 \leq t \land t1 \in LE(e)) \\
inv4 : change \in \mathbb{N}
\end{array}
$$

Initially, the system time is initialized to zero ($t = 0$). Then, the edges which are present at the time "$t = 0$" have the label "0" ($LE=\{0\}$).

If we say that an edge "e" exists in "g" ($e \in g$), then its label contains at least one date "$t1$" knowing that $t1 \leq t$ (see inv3).

As we said above, the incrementation of time, from a date "t" to a date "$t+1$", is done after one or several topological events (adding edge, removing edge). To do so, we introduce a new variable of type integer called "$change$" (see inv4). This variable is initialized to zero ($change=0$) and if an event takes place, then "$change=1$".

In order to express the different topological changes over time, based on the evolving graphs, we explain and we specify the different events as follows:

- **Adding_Edge:** By introducing the concept of time, the appearance of a new edge in the network requires the addition of the edge and a label containing the current date "t". Thus, the "*Adding_Edge*" event presented in *M0* is refined by modifying the action component. In fact, we add the new edge and then we add a label containing the current date. Also, the variable "$change$" receives the value "1" (act2) to indicate that a topological change has been produced. We provide below the specification of the "*Adding_Edge*" event.

- **Maintaining_Edge:** If an edge has appeared at a date "$t1$", with "$t1$" strictly lower than the current date "t", and it still exists at the date "t", then we call this event "*Maintaining_Edge*". Formally, in the guard component, we define "grd1" to verify the existence of an edge "e" in the graph "g" before the date "t". Also, we add "grd2" to guarantee that the date "t" does not belong to the label of the edge "e". In the action component, we update the label of the edge "e" by adding the date "t" to the existing label (act1). The variable "$change$" does not change since the network has not undergone any modification. The "*Maintaining_Edge*" event is specified as follows:

```
Adding_Edge
refines Adding_Edge
any   x, y
where
  grd1 : x ↦ y ∈ V × V
  grd2 : x ↦ y ∉ g
  grd3 : x ≠ y
then
  act1 : g, LE : |g' = g ∪ {x ↦ y, y ↦ x}
     ∧LE' = LE ∪ {(x ↦ y) ↦ {t}, (y ↦ x) ↦ {t}}
  act2 : change := 1
end
```

```
Maintaining_Edge
any   e
where
  grd1 : e ∈ g
  grd2 : t ∉ LE(e)
then
  act1 : LE(e) := LE(e) ∪ {t}
end
```

- **Removing_Edge:** An edge has been removed at the actual date *"t"* if it exists at the date *"t-1"* (see grd1 and grd2), but it does not exist at the date *"t"* (see grd3). In this situation, nothing will change in the evolving graph. So, in the action component, we modify only the variable *"change"* (act1). We provide the specification of the *"Removing_Edge"* event below.
- **Incrementing_Time:** We have introduced a new event, called *"Incrementing_Time"*, which ensure the incrementation of time. In the guard component, we verify that $t \geq 0$ and $t < tn$ (grd1). This event is activated when one or several topological events (*Adding_Edge, Removing_Edge*) occur in the network, which means that *"change"* is equal to 1 (grd2). In the action component, we increment the time to *"t +1"* and we reset the variable *"change"* (act2). Then, we have no topological change at the time *"t +1"*. The *"Incrementing_Time"* specification is done as follows:

```
Removing_Edge
any   e
where
  grd1 : e ∈ g
  grd2 : t ≥ 1 ∧ (t − 1) ∈ LE(e)
  grd3 : t ∉ LE(e)
then
  act1 : change := 1
end
```

```
Incrementing_Time
where
  grd1 : t ≥ 1 ∧ t < tn
  grd2 : change = 1
then
  act1 : t := t + 1
  act2 : change := 0
end
```

Using the proposed model, it is possible to find the historical data of a dynamic network. So, it is a way to analyze an evolving graph. In fact, through the function *"LE"*, we can obtain the presence dates of each edge in the network. Then, we can verify the connectivity over time [7] of the network. Also, we can find the set of all possible paths over time from one node to another, called journeys. Thus, it is possible to compute optimal journeys in dynamic networks like the minimum delay of path (fastest journey), the earliest arrival date (foremost journey) and the minimum number of hops (shortest journey) [5].

5 Example

In this section, we present an illustration of our model by specifying a distributed algorithm encoded by local computations model. For this purpose, we choose a simple example of algorithm, called centralized counting algorithm. It is a distributed

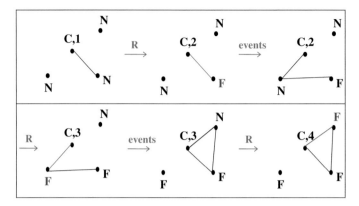

Fig. 6 Execution example of the centralized counting algorithm over evolving graph

algorithm which computes all the nodes in a network. In fact, each entity executes asynchronously the same code and interacts locally with its immediate neighbours. The proposed model can be applied to other distributed algorithms. The main objective of this section is to demonstrate how our model can be used and incorporated during development. Through this example, we list and we discuss the instantiation of the different model components, by refinement technique, to generate a correct specification. We begin in this section by presenting the chosen algorithm. After that, we illustrate how we obtain an instance of the proposed model.

5.1 Algorithm Presentation

The centralized counting algorithm, depicted in Algorithm 1, assumes a distinguished node at initial time. This node, called the counter, is in charge of counting all the nodes it meets during the execution. Therefore, the counter node has two labels (C, i), meaning that it is the counter (C), and that it has already counted i nodes (initially 1, i.e., itself). The other nodes are labeled either "F" or "N", depending on whether they have already been counted or not. The counting rule is given by the relabelling rule "R" in Algorithm 1. An execution example of this algorithm over the evolving graph is given in Fig. 6.

Algorithm 1 Counting algorithm with pre-selected counter.

initial states: $(C,1),N$ $((C,1)$ for the counter, N for all other nodes)
alphabet: C, N, F, $\mathbb{N}^*$
rule R:

$$R: \quad \overset{C,i}{\bullet} \xrightarrow{\quad N \quad} \overset{C,i+1}{\bullet} \xrightarrow{\qquad} \overset{F}{\bullet}$$

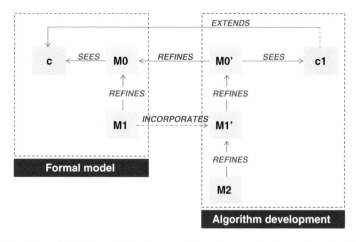

Fig. 7 Using the model in Event-B development (Example: centralized counting algorithm)

5.2 Formal Specification

To explain the specification of the centralized counting algorithm, we present in Fig. 7 how the proposed model is used to construct a correct distributed algorithm. In fact, we begin by specifying the application field of the algorithm. Then, we introduce a new context called "c1", as an extension of the context "c", which describes the specific algorithmic properties. Generally, the development of a distributed algorithm starts with a very abstract algorithm and then by successive refinement we obtain a concrete one that expresses the local behavior of the processor in the network.

According to this development strategy and with respect to the given algorithm, three basic levels are necessary to build a correct distributed algorithm.

In the first level, we define a machine *M0'* which refines *M0*. Then, it includes the events *"Adding_Edge"* and *"Maintaining_Edge"*. *M0'* can access to all components of the context *"c1"* (via the clause *SEES*). It expresses only the goal of the distributed algorithm through a new event *"oneshot"* which does not describe how the solution is computed. **In the second level**, *M1* is incorporated into *M1'* (via the clause *INCORPORATES*). Thus, *M1'* includes the events *"Adding_Edge"*, *"Maintaining_Edge"*, *"Removing_Edge"* and *"Incrementing_Time"* as defined in *M1*. In addition, we refine the events *"Adding_Edge"* and *"Maintaining_Edge"* of *M0'* in the same manner as the refinement done in the proposed model. Also, we refine the event *"oneshot"*. **In the third level**, we introduce a machine, called *M2*, which refines *M1'*. In this level, we specify the local label modification and encode the relabelling rule described above.

The details of the context and machines development will be explained afterwards.

- **The context *"c1"***

  In this context, we define the node "c0" as a constant which is in charge of counting all the nodes it meets during the execution. We denote node labels by a set called

"LN": The node "c0" is labeled "C". The other nodes are labeled "F" if they have already been counted and "N" if they have not (see axm 2). All these specific algorithmic properties are specified as follows:

```
context c1
extends c
sets
   LN
constants
   c0, N, F, C
axioms
   axm1 : c0 ∈ V
   axm2 : partition(LN, {C}, {N}, {F})
end
```

- **The first level: Machine *M0'***
  In this level, the *"Adding_Edge"* and the *"Maintaining_Edge"* events remain unchanged and we add an event called *"oneshot"*. This event avows the result of the distributed algorithm when its execution is completed. Then, it returns the number of nodes in the network without describing how the solution is computed. In order to specify the *"oneshot"* event, we introduce two variables *"nodes"* and *"nb_nodes"* which will contain respectively the sets of nodes that have been counted and the resulting number of nodes. These new variables are specified by the following invariants:

```
inv4 : nodes ⊆ V
inv5 : nodes ≠ ∅
inv6 : c0 ∈ nodes
inv7 : ∀x · x ∈ nodes\{c0} ⟹ c0 ↦ x ∈ g
inv8 : nb_nodes ∈ ℕ
```

Initially, *"nodes"* contains the center "c0". Thus, *"nb_nodes"* is equal to 1. The following initialization establishes the invariants:

```
act1 : g : |(g' ⊆ V × V) ∧ (g' = g'⁻¹) ∧ ((V ◁ id) ∩ g' = ∅)
act2 : nodes := {c0}
act3 : nb_nodes := 1
```

We provide the specification of the *"oneshot"* event below. In the guard component, all nodes, except "c0", must be connected to the center "c0". In the action component, the variable *"nodes"* contains all the network nodes and *"nb_nodes"* is the number of nodes in the network. We prove by means of the theorem "Th1" that the graph *"g"* is connected.

```
oneshot
where
   grd1 : ∀x · x ∈ V \ {c0} ⟹ c0 ↦ x ∈ g
   Th1 : ∀s · s ⊆ V ∧ s ≠ ∅ ∧ g[s] ⊆ s ⟹ V ⊆ s
then
   act1 : nodes := V
   act2 : nb_nodes := card(V)
end
```

- **The second level: Machine *M1'***

  The refinement of *M0'*, named *M1'*, remains in a high level abstraction. It encodes the algorithm and computes its result without considering the relabelling rules. The specification presented in the second level of the model still exists. However, we have to refine the *"oneshot"* event defined in *M0'*, due to the presence of the time aspect in this level. In fact, the "grd1" presented in *M0'* is reinforced by a new condition called "grd2". It ensures that each node "x" in the graph, except "c0", will be linked to the node "c0" at one or several dates. Then, the label of each edge "$c0 \mapsto x$" in the graph contains one or several dates. The guard "grd2" of the *"oneshot"* event specification is given as follows:

  $$grd2 : \forall x \cdot x \in V \setminus \{c0\} \implies (\exists t1 \cdot t1 \geq 0 \wedge t1 \leq tn \wedge t1 \in LE(c0 \mapsto x))$$

- **The third level: Machine *M2***

  The third machine, called *M2*, refines the previous one. It introduces labels of nodes and edges. The *"oneshot"* event still exists but it is more concrete. However, the other events (*Adding_Edge, Maintaining_Edge, Removing_Edge*), presented in *M1'*, remain unchanged.

  Let "lab" be the variable which describes the states of nodes: "$lab \in V \to LN$" (inv1) where "LN" is defined in the context "c1" as the set of possible labels for the nodes. At every time, each node is in a particular state and this state will be encoded by a node label. According to its own state and to the states of its neighbours, each node may decide to perform a computation step by applying the relabelling rule. Initially, the node "c0" is labeled "C" and the other nodes are labeled "N", since they have not been counted. The initialization of the variable *"lab"* is defined as follows:

  $$act5 : lab := ((V \setminus \{c0\}) \times \{N\}) \cup (\{c0 \mapsto C\})$$

  Formally, the relabelling rule "R" is specified by the event *"Rule"* as follows:

```
Rule
any
    s2
where
    grd1 : c0 ↦ s2 ∈ g
    grd2 : lab[{s2}] = {N}
    grd3 : t ∈ LE(c0 ↦ s2)
then
    act1 : lab(s2) := F
end
```

  We also define a gluing invariant called *"gluing_inv"*. Generally, the gluing invariants are used to relate the states between the concrete and abstract machines. In our context, if a node "x" belongs to "nodes", which represents the sets of nodes that have been counted, then it is connected to the center "c0" and labeled "F". Furthermore, the edge joining the node "x" and "c0" should be present in one or several times. The gluing invariant specification of *M2* is done as follows:

$$gluing\_inv : \forall x \cdot x \in nodes \setminus \{c0\} \implies lab(x) = F$$
$$\wedge c0 \mapsto x \in g \wedge (\exists t1 \cdot t1 \geq 0 \wedge t1 \leq tn \wedge t1 \in LE(c0 \mapsto x))$$

In this level, we refine the *"oneshot"* event by adding new guards and reinforcing the action "act1" of the abstract event. In fact, we reinforce the guard component by adding "grd3", "grd4" and "grd5" which ensure that the node "c0" is labeled "C" (grd3), all the other nodes are labeled "F" (grd4) and no node is labeled "N" (grd5). The added guards in the *"oneshot"* event specification is given as follows:

$$grd3 : lab[\{c0\}] = \{C\}$$
$$grd4 : lab[V \setminus \{c0\}] = \{F\}$$
$$grd5 : lab^{-1}[\{N\}] = \emptyset$$

Also, we reinforce the action "act1" to indicate that, at the end of the execution of the algorithm, the nodes are labeled "C" or "F".

$$act1 : nodes := lab \sim [\{C, F\}]$$

We prove by means of the theorem 2 (Th2) that our algorithm can apply its rewriting rules to every edge before "tn".

$$Th2 : \forall x \cdot x \in V \setminus \{c0\} \wedge c0 \mapsto x \in g \implies (\exists t1 \cdot t1 \geq 0 \wedge t1 \leq tn \wedge t1 \in LE(c0 \mapsto x) \wedge lab(x) = F)$$

6 Conclusion and Future Work

In this paper, we have presented a formal and general model for dynamic networks based on the evolving graph formalism. It aims to define the different topological changes and the situations of time evolution. It is also a way to analyze the evolving graph. The proposed model is based on the refinement technique by using the Event-B formal method and the RODIN platform. The main characteristic of this model is that it enables reuse in the development and minimizes efforts on proving distributed algorithms. We have illustrated it by investigating an example of the centralized counting algorithm.

We are currently working on dealing with other examples of distributed algorithms which are more complex. We plan to extend our model by introducing some properties related to evolving graphs such as connectivity over time, journeys, etc. Moreover, it is interesting to choose a case study supporting the dynamic behavior of the network in order to apply the proposed model in realistic scenarios.

References

1. Abrial, J.: Modeling in Event-B–System and Software Engineering. Cambridge University Press, Cambridge (2010)
2. Abrial, J.R., Butler, M., Hallerstede, S., Hoang, T., Mehta, F., Voisin, L.: Rodin: an open toolset for modelling and reasoning in event-b. Int. J. Softw. Tools Technol. Transf. **12**(6), 447–466 (2010)
3. Back, R.J.R.: A calculus of refinements for program derivations. Acta Informatica **25**, 593–624 (1988)
4. Barjon, M., Casteigts, A., Chaumette, S., Johnen, C., Neggaz, Y.M.: Maintaining a spanning forest in highly dynamic networks: the synchronous case. In: Proceedings of the Principles of Distributed Systems–18th International Conference, OPODIS 2014, Cortina d'Ampezzo, Italy, 16-19 December 2014, pp. 277-292 (2014)
5. Bui-Xuan, B.M., Ferreira, A., Jarry, A.: Computing shortest, fastest, and foremost journeys in dynamic networks. Int. J. Found. Comput. Sci. **14**, 267–285 (2003)
6. Casteigts, A.: Contribution à l'algorithmique distribué dans les réseaux mobiles ad hoc. Ph.D. thesis, Université Sciences et Technologies–Bordeaux I (2007)
7. Casteigts, A., Chaumette, S., Ferreira, A.: Distributed Computing in Dynamic Networks: Towards a Framework for Automated Analysis of Algorithms. CoRR, abs/1102.5529 (2012)
8. Ferreira, A.: Building a reference combinatorial model for MANETs. IEEE Network **18**(5), 24–29 (2004)
9. Floriano, P., Goldman, A., Arantes, L.: Formalization of the necessary and sufficient connectivity conditions to the distributed mutual exclusion problem in dynamic networks. In: Proceedings of the 2011 IEEE 10th International Symposium on Network Computing and Applications, NCA'11, pp. 203-210. IEEE Computer Society, Washington (2011)
10. Kerchove, F.M.D.: Relabeling Algorithms on Dynamic Graphs. Technical report, University of Le Havre (2012)
11. Leavens, G.T., Abrial, J.R., Batory, D., Butler, M., Coglio, A., Fisler, K., Hehner, E., Jones, C., Miller, D., Peyton-Jones, S., Sitaraman, M., Smith, D.R., Stump, A.: Roadmap for enhanced languages and methods to aid verification. In: Proceedings of the 5th International Conference on Generative Programming and Component Engineering, GPCE'06, pp. 221-236. ACM, New York (2006)
12. Litovsky, I., Métivier, Y., Sopena, E.: Handbook of graph grammars and computing by graph transformation. In: Graph Relabelling Systems and Distributed Algorithms, pp. 1-56. World Scientific Publishing Co., Inc., River Edge (1999)
13. Roy, R.: Mobile ad hoc networks. In: Handbook of Mobile Ad Hoc Networks for Mobility Models, pp. 3-22. Springer, New York (2011)
14. Tel, G.: Introduction to Distributed Algorithms, 2nd edn. Cambridge University Press, New York (2001)

An Iterated Variable Neighborhood Descent Hyperheuristic for the Quadratic Multiple Knapsack Problem

Takwa Tlili, Hiba Yahyaoui and Saoussen Krichen

Abstract The Quadratic Multiple Knapsack Problem (QMKP) is a variant of the well-known NP-hard knapsack problem that assign profits not only to individual items but also to pairs of items. QMKP aims to maximize a quadratic objective function subject to a linear capacity constraint. In this paper, we focus on proposing a hyper-heuristic approach based in the iterated variable neighborhood descent algorithm for solving the QMKP. Numerical investigations based on well-known benchmark instances are conducted. The results clearly demonstrate the good performance of the proposed algorithm in solving the QMKP.

Keywords Hyper-heuristic · Iterated variable neighborhood descent · Quadratic multiple knapsack problem

1 Introduction

The knapsack problem (KP), a well-known NP-hard optimization problem, has been thoroughly studied in the literature. KP can be encountered in numerous real-world applications in different areas, such as, resource allocation [3], investment decision-making [8], network interdiction problem, location problems [12], and capital budgeting problems [1]. The standard KP is defined as follows. Given a set of items with profits and weights, and a single knapsack, the aim is to maximize the total profits of selected items to put into the knapsack subject to the capacity constraint.

The Quadratic Knapsack Problem (QKP) firstly evoked by Gallo et al. [4] is a generalization of the classical KP, which the total profit should reflect how well the selected items fit together. A more recent extension of the QKP considers numerous knapsacks, called the Quadratic Multiple Knapsack Problem (QMKP), where profit values are assigned not only to the items but to pairs of them.

T. Tlili (✉) · H. Yahyaoui · S. Krichen
LARODEC, Institut Supérieur de Gestion Tunis, Université de Tunis, Tunis, Tunisia
e-mail: takwa.tlili@gmail.com

© Springer International Publishing Switzerland 2016
R. Lee (ed.), *Software Engineering, Artificial Intelligence, Networking and Parallel/Distributed Computing 2015*, Studies in Computational Intelligence 612,
DOI 10.1007/978-3-319-23509-7_17

245

In this paper, we handle the multiple quadratic knapsack problem using a new Hybridization between Hyperheuristic-based Variable Neighborhood Search and the iterated local search (HH-VNS).

The rest of the paper is organized as follows. Section 2 provides a detailed description of the problem. Section 3 describes the proposed iterated variable neighborhood descent hyper-heuristic for the MQKP presents. Section 4 presents the numerical results compared to the other state-of-art approaches. Finally we conclude in Sect. 5.

2 Quadratic Multiple Knapsack Problem

The quadratic multiple knapsack problem (QMKP) is about packing a set of items disjunctively to a set of knapsacks while maximising the total profit subject to the capacity constraint. Formally the QMKP can be described as follows.

Formally, the QMKP is to assign n items to m knapsacks such that the total profit of the allocated items is maximized. Each knapsack $k \in \{1, 2, \ldots, m\}$ has a maximum capacity C_k. Each object $i \in \{1, 2, \ldots, n\}$ has a profit pi and a size si. Each pair of items (i, j) is associated with a joint profit p_{ij}. QMKP seeks to assign each item to a single knapsack such that the total weight of the packed items does not exceed the capacity of used knapsack. Maximizing the total profit of all the assigned items is our objective.

The QMKP can be formulated as follows [2, 5, 6].

$$Max \quad Z(X) = \sum_{i=1}^{n} \sum_{k=1}^{m} x_{ik} p_i + \sum_{i=1}^{n-1} \sum_{j=i+1}^{n} \sum_{k=1}^{m} x_{ik} x_{jk} p_{ij} \tag{1}$$

Subject to

$$\sum_{i=1}^{n} x_{ik} w_i \leq C_k; \forall k \tag{2}$$

$$\sum_{k=1}^{m} x_{ik} \leq 1; \forall i \tag{3}$$

where the variables xik indicate that item i is included in knapsack k.

Hiley and Julstrom [7] proposed the first study about the QMKP which is a combination of two optimization problems: the multiple-knapsack and QKP. Authors proposed three different heuristics: a greedy heuristic, a stochastic hill-climbing method and a genetic algorithm. Since then, a variety of metaheuristics have been developed for the QMKP. Saraç and Sipahioglu [9] proposed a genetic-based algorithm as well as a hybrid approach for the QMKP. García-Martínez et al. [5, 6] proposed a tabu-enhanced destruction mechanism for iterated greedy search, named the tabu-enhanced iterated greedy algorithm. Authors proved that the resulting algorithm exploits effectively the problem-knowledge associated with the QMKP requirements. Sundar and Singh [11] have incorporated group preserving property into Artificial Bee Colony (ABC) method for QMKP. García-Martínez et al. [5, 6] developed a strategic oscillation based framework that operates in relation to a critical boundary associated with important solution features. Authors showed that their proposed approach outperforms current state of-the-art algorithms. Soak et al. [10] proposed a memetic algorithm which is based on the adaptive link adjustment evolutionary algorithm (ALA-EA). More recently, Chen and Hao [2] developed a new algorithm named, iterated responsive threshold search (IRTS) approach for solving the QMKP. The proposed IRTS discovers 41 improved lower bounds and reached all the best known results for the benchmark instances.

3 Iterated Variable Neighborhood Descent Hyper-heuristic for the QMKP

Hyper-heuristic refers to the coupling of artificial intelligence methods with automated theorem proving. It is a high-level search strategy that either generates or selects a heuristic from a number of low-level heuristics at each point in the search. In this paper, we developed the Iterated Variable Neighborhood Descent Hyperheuristic (IVND-H) expressed as shown in Fig. 1.

The framework resolution depicted in Fig. 1 can be formally detailed as follows. IVND-H starts from an initial solution and iteratively explores its neighborhood looking for a better one. Each integrated heuristic is associated with a weight which is updated according to the performance during the search process. IVND-H iterates until no further improvements is obtained.

4 Experimental Results

This section is devoted to describe the preliminary computational experiments carried out to assess the quality of the solutions provided by the IVND-H for the QMKP. The proposed approach was built in Java and ran on a personal computer with 2.4 GHz $Intel^{®}$ $Core^{TM}$ processor, 4 GB RAM and Windows 7 as an operating system. In order to show the effectiveness of the IVND-H, we apply it on a benchmark instances available on line at http://www.optsicom.es/qmkp/. The data set contains 60 instances in which the number of items $n \in \{100, 200\}$ and the number of knapsacks $m \in \{3, 5, 10\}$. Hiley and Julstrom [7] used the first five instances from each group

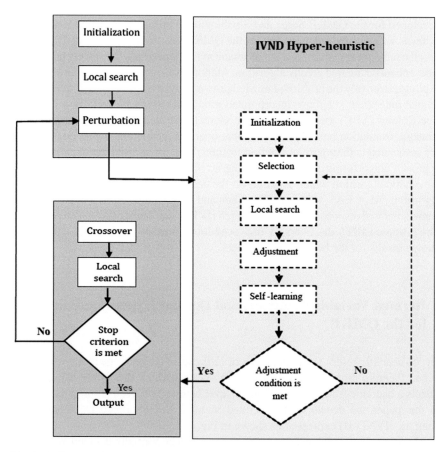

Fig. 1 IVND hyper-heuristic for the QMKP

of density 0.25 to generate the first set of thirty QMKP instances. The other set of instances were generated in the same way from the posted QKP instances with density 0.75. For each QMKP instance, knapsack capacities are set to 80 % of the sum of the objects weights divided by the number of knapsacks.

In Table 1, we present the empirical results of both the IVND and IVND-H. Columns in Table 1 are respectively number of items n, number of knapsacks m, N^o of instance I, knapsack capacity C, solution of VND-H algorithm and solution of IVND-H. Table 1 shows that in almost all instances the IVND-H outperforms the VND-H due to the perturbation that provide better investigation of the search space.

Table 1 Empirical results for the QMKP (with density 25 and 75 %) given by IVND-H, IVND and the Imp (%)

Instance								Instance						
n	m	l	C	VND-H	IVND-H	IM (%)		n	m	l	C	VND-H	IVND-H	IM (%)
100	3	1	688	24720	25369	2.5		100	3	1	669	49341	51731	4.6
100	3	2	738	25675	26398	2.7		100	3	2	714	51510	52464	1.8
100	3	3	663	25225	26752	5.7		100	3	3	686	49526	51724	4.2
100	3	4	804	26314	27839	5.4		100	3	4	666	51403	51452	0.1
100	3	5	723	25193	26835	6.1		100	3	5	668	45924	51218	10.3
100	5	1	413	20358	21694	6.1		100	5	1	401	47568	48865	2.6
100	5	2	442	19156	20236	5.3		100	5	2	428	47259	48125	1.8
100	5	3	398	19232	20935	8.1		100	5	3	411	46985	47562	1.2
100	5	4	482	20896	21453	2.5		100	5	4	400	48236	49275	2.1
100	5	5	434	19567	20986	6.7		100	5	5	400	46632	47862	2.5
100	10	1	206	14562	15268	4.6		100	10	1	200	27682	28429	2.6
100	10	2	221	13214	14965	11.7		100	10	2	214	28731	29765	3.4
100	10	3	199	13025	13954	6.6		100	10	3	205	27946	28761	2.8
100	10	4	241	14526	15439	5.9		100	10	4	200	29765	30948	3.8
100	10	5	217	13625	14875	8.4		100	10	5	200	28765	29145	1.3
200	3	1	1381	89562	90583	1.12		200	3	1	1311	268068	269458	0.5
200	3	2	1246	89421	90763	1.4		200	3	2	1414	255761	256143	0.1
200	3	3	1246	89473	90714	1.3		200	3	3	1342	268479	269431	0.3
200	3	4	1413	95345	97143	1.8		200	3	4	1565	244135	245679	0.6
200	3	5	1358	98376	99238	0.8		200	3	5	1336	276318	278456	0.7

(continued)

Table 1 (continued)

Instance				VND-H	IVND-H	IM (%)	Instance				VND-H	IVND-H	IM (%)
n	m	I	C				n	m	I	C			
200	5	1	828	72912	73845	1.2	200	5	1	786	182364	184695	1.2
200	5	2	747	76953	78142	1.5	200	5	2	848	170698	173268	1.4
200	5	3	801	75432	76354	1.2	200	5	3	805	184690	185247	0.3
200	5	4	848	71486	72678	1.6	200	5	4	939	163984	165876	1.1
200	5	5	815	74832	75246	0.5	200	5	5	801	191986	192658	0.3
200	10	1	414	49287	50138	1.6	200	10	1	393	110863	111236	0.3
200	10	2	373	52476	53268	1.4	200	10	2	424	102987	103684	0.6
200	10	3	400	49862	51027	2.2	200	10	3	402	111536	112948	1.2
200	10	4	424	47965	49237	2.5	200	10	4	469	96947	97543	0.6
200	10	5	407	50632	51258	1.2	200	10	5	400	114863	115369	0.4

5 Conclusion

In this paper, we addressed the quadratic multiple knapsack problem (QMKP) which is a knapsack problem with multiple knapsacks and items joint profits. For handling the QMKP, we developed the iterated variable neighborhood descent hyper-heuristic (IVND-H). The proposed algorithm does not impose a heuristic but the selection is done according to the score of each heuristic. The conducted numerical investigations clearly demonstrated the good performance of the proposed algorithm to solve the QMKP.

References

1. Bas, E.: A capital budgeting problem for preventing workplace mobbing by using analytic hierarchy process and fuzzy 0–1 bidimensional knapsack model. Expert Syst. Appl. **38**, 12415–12422 (2011)
2. Chen, Y., Hao, J.-K.: Iterated responsive threshold search for the quadratic multiple knapsack problem. Ann. Oper. Res. **226**, 101–131 (2015)
3. Deng, Y., Chen, Y., Zhang, Y., Mahadevan, S.: Fuzzy dijkstra algorithm for shortest path problem under uncertain environment. Appl. Soft Comput. **12**, 1231–1237 (2011)
4. Gallo, G., Hammer, P., Simeone, B.: Quadratic knapsack problems. Math. Program. **12**, 132–149 (1980)
5. Garcia-Martinez, C., Rodriguez, F.J., Lozano, M.: Tabu-enhanced iterated greedy algorithm: a case study in the quadratic multiple knapsack problem. Eur. J. Oper. Res. **232**, 454–463 (2014)
6. García-Martínez, C., Glover, F., Rodriguez, F.J., Lozano, M., Martí, R.: Strategic oscillation for the quadratic multiple knapsack problem. Comput. Optim. Appl. **58**, 161–185 (2014)
7. Hiley, A., Julstrom, B.A.: The quadratic multiple knapsack problem and three heuristic approaches to it. Proc. Genet. Evol. Comput. Conf. **1**, 547–552 (2006)
8. Peeta, S., Salman, F., Gunnec, D., Viswanath, K.: Pre-disaster investment decisions for strengthening a highway network. Comput. Oper. Res. **37**, 1708–1719 (2010)
9. Saraç, T., Sipahioglu, A.: Generalized quadratic multiple knapsack problem and two solution approaches. Comput. Oper. Res. **43**, 78–89 (2014)
10. Soak, S.-M., Lee, S.-W.: A memetic algorithm for the quadratic multiple container packing problem. Appl. Intell. **36**, 119–135 (2012)
11. Sundar, S., Singh, A.: A swarm intelligent approach to the quadratic multiple knapsack problem. In: LNCS, pp. 626–633. Springer (2010)
12. Vanderster, D., Dimopoulos, N., Parra-Hernandez, R., Sobie, R.: Resource allocation on computational grids using a utility model and the knapsack problem. Future Gener. Comput. Syst. **25**, 35–50 (2009)

Author Index

© Springer International Publishing Switzerland 2016
R. Lee (ed.), *Software Engineering, Artificial Intelligence, Networking
and Parallel/Distributed Computing 2015*, Studies in Computational Intelligence 612,
DOI 10.1007/978-3-319-23509-7